Electricity for Air Conditioning and Refrigeration Technicians

Order of course

Review – Unit 8
- Unit 10 (in depth)
- Unit 11 (in depth)
- Unit 12 (trade related)
- Unit 13 (new)
- Unit 14 (handouts) given
- Unit 15 (brief)
- Unit 16. Control devices.
- Unit 18 AC circuits
- Unit 19 Refrigeration
- Unit 20 Troubleshooting
- Unit 21 Gas furnaces
- Unit 22 – delivery systems for conditioned air

* – Appendix A

* – Glossary

Fourth Edition

Electricity for Air Conditioning and Refrigeration Technicians

Edward F. Mahoney

PRENTICE HALL
Upper Saddle River, New Jersey   Columbus, Ohio

Library of Congress Cataloging-in-Publication Data

Mahoney, Edward F.
  Electricity for air conditioning and refrigeration technicians /
Edward F. Mahoney. — 4th ed.
      p.   cm.
  Includes index.
  ISBN 0-13-371642-2
    1. Air conditioning—Electric equipment.   2. Refrigeration and
refrigerating machinery—Electric equipment   I.Title.
  TK4035.A35M33   1997
  697.9'3—dc20   96-31481
                                                                 CIP

Acquisitions Editor: Ed Francis
Production Editor: Alex Wolf
Cover Design: Russ Maselli
Production Manager: Deidra Schwartz
Editorial-Production Supervision: WordCrafters Editorial Services, Inc.
Marketing Manager: Danny Hoyt

This book was set in Meridien by BookMasters and was printed and bound by R.R.
Donnelley & Sons, Inc. The cover was printed by Phoenix Color Corp.

 © 1997, 1993, 1986, 1980 by Prentice-Hall, Inc.
Simon & Schuster/A Viacom Company
Upper Saddle River, New Jersey 07458

Printed in the United States of America

10 9 8 7 6 5 4 3 2 1

ISBN 0-13-371642-2

Prentice-Hall International (UK) Limited, *London*
Prentice-Hall of Australia Pty. Limited, *Sydney*
Prentice-Hall of Canada, Inc., *Toronto*
Prentice-Hall Hispanoamericana, S.A., *Mexico*
Prentice-Hall of India Private Limited, *New Delhi*
Prentice-Hall of Japan, Inc., *Tokyo*
Simon & Schuster Asia Pte. Ltd., *Singapore*
Editora Prentice-Hall do Brasil, Ltda., *Rio de Janeiro*

*To my wife, Gloria, whose assistance made this book possible
and to my mother, Clara, the original technician*

Preface

Many changes have taken place in the control of air-conditioning systems since the original publication of this book in 1980. High-efficiency condensing units are standard today. Both single- and three-phase two-speed motors are being used to drive compressors. Variable-speed three-phase motors are beginning to show promise in the condensing units of commercial systems and are providing increased efficiency.

It is becoming increasingly important that today's air-conditioning and refrigeration technicians have a well-founded background in electricity and also be familiar with electronic concepts.

Important revisions in this edition include the addition of Objectives and Practical Experience exercises to each unit.

Thermistors and the semiconductor sensing element have been added to Unit 17, as have discussions of both the effect of low voltage on a compressor operation and of the low-voltage–anti-short-cycle cutout.

Two new units have been added to this edition. Unit 21 covers gas-furnace controls, while Unit 22 covers conditioned-air delivery systems. Unit 22 also provides basic information on electronic air cleaners.

The beginning of study in a different area of technology is cause for concern because of the new terms that must be mastered. In the area of air conditioning and refrigeration, there are a number of control devices peculiar to the field. The names of these devices and their symbols as used in diagrams is important.

Therefore, it is suggested that before starting this textbook, the reader examine Appendix A, where the symbols for many of the devices are presented, and Appendix B, Powers of Ten, for coverage of how to work with large numbers. The symbols are repeated on the inside covers and the last three pages of the book.

Brief Contents

Contents

Unit 12 TRANSFORMERS 122

Unit 13 PHASE SHIFT AND POWER FACTOR 137

Unit 14 ELECTRIC MOTORS 150

Unit 15 MOTOR-STARTING CIRCUITS 181

Electron Theory

OBJECTIVES

After study and review of this unit, you will be able to
• Relate electricity to those things that are common in your environment.
• Picture in your mind electrons and electrons moving.
• Define the terms *voltage* and *current.*
• Recognize the difference between an open and a closed circuit.

The information presented in this unit is theoretical. Theory is a line of reasoning that is assumed to be correct. Theory may change from time to time as new information is gathered and new lines of reasoning are formed.

Electrical theory is highly developed at the present time and requires a knowledge of higher mathematics for complete understanding. However, mathematical explanations of electricity are not the concern of this text; instead, an elementary approach to those electrical concepts, theories, and formulas that are essential to the air-conditioning and refrigeration technician will be provided.

It is suggested that this unit be read as a story. The material presented is good background information that will assist the reader in understanding the nature of electricity. With a good grasp of the nature of electricity, an air-conditioning and refrigeration technician will find troubleshooting an easier task.

THE STRUCTURE OF MATTER

A negatively charged, tiny object called an *electron* must be considered in order to understand the behavior of electricity. This consideration requires an investigation into the structure of matter—of what are things really made?

Imagine looking at a piece of copper under a microscope as the magnification is increased to an extremely high magnitude. The actual construction of copper might be seen.

With the microscope at 10× power, the copper looks very much like copper (Figure 1-1). As the magnification is increased to 100×, the rough crystalline structure of copper is seen (Figure 1-2). It takes a large jump in magnification to 10,000,000× before the beginning of atomic structure becomes evident, as seen in the bumpy surface (Figure 1-3). Finally, at a magnification of 100,000,000×, individual atoms are seen (Figure 1-4).

What is there to see? The center of a copper atom, the *nucleus*, consists of positively charged and neutral particles, *protons* and *neutrons*. Negatively charged particles, *electrons,* whirl in four orbits around the center. Electron theory is based on these negatively charged particles. An accepted representation of a copper atom is given in Figure 1-5.

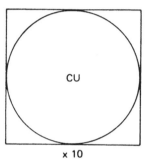

× 10

FIGURE 1-1   Plain copper (Cu).

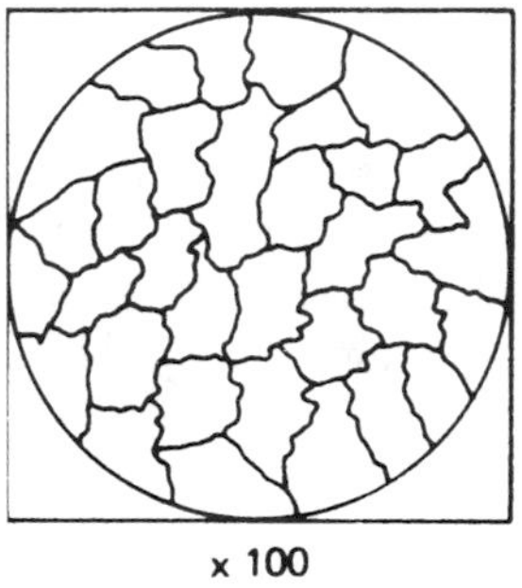

× 100

FIGURE 1-2   Copper (Cu) crystal.

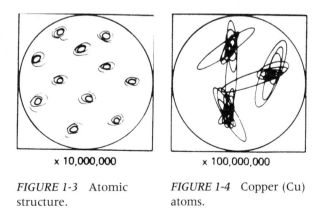

× 10,000,000

FIGURE 1-3   Atomic structure.

× 100,000,000

FIGURE 1-4   Copper (Cu) atoms.

In the study of *electricity,* the electrons whirling about the nucleus are the center of interest. Moving these electrons on command produces electricity that can be used to do work.

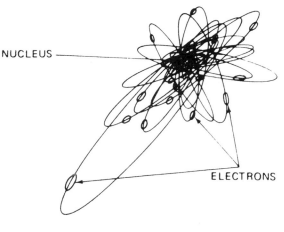

NUCLEUS

ELECTRONS

FIGURE 1-5   COPPER ATOM

BOHR'S LAW

As scientists experimented with things of an electrical nature, they learned about the characteristics of electricity and about electrical phenomena. Physicist Niels H. D. Bohr (1885–1962) developed a model of electron, proton, and neutron arrangements to represent everything in the universe; this became known as Bohr's quantum theory of atomic structure. According to Bohr, an atom of hydrogen, the lightest known element, consists of one proton in the nucleus and one electron rotating about it (Figure 1-6). Bohr's theory of the hydrogen atom is regarded as the basis of modern atomic nuclear physics.

FIGURE 1-6   HYDROGEN ATOM

As we move up the atomic scale, heavier materials have a greater number of particles in the nucleus and a greater number of electrons whirling about that center. As the theory developed and is now understood, the arrangement of electrons in rings (orbits) around the nucleus is fixed.

1. In the first orbit around the nucleus, there can be no more than 2 electrons.

2. In the second orbit, the maximum is 8 electrons.

3. The third orbit has a maximum of 18 electrons.

4. The fourth orbit has a maximum of 32 electrons.

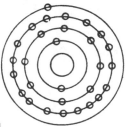

FIGURE 1-7   Electron configuration
of copper atom.

COPPER ATOM

CONDUCTORS

As indicated, the number of electrons in each orbit around the nucleus of an atom is fixed. The number of electrons in orbit and the number of protons at the center determines the type of material.

Copper, for example, has a nucleus of 29 protons and 34 neutrons with 29 electrons rotating around it. According to the fixed arrangement of electrons, there are 2 in the first orbit, 8 in the second orbit, and 18 in the third orbit. This accounts for only 28 electrons. The remaining electron orbits all by itself in the fourth ring (see Figure 1-7). Since this electron is all by itself in the fourth ring, it is not held as strongly to the nucleus as are the other electrons. The electron is "free" to move whenever an outside force acts on it. Copper is a good conductor of electricity because each of the billions of tiny atoms in a copper conductor has one free electron ready to move when a force is applied to it. A reasonably small copper wire will pass a relatively large amount of electricity.

Another example of a good conductor is aluminum. The aluminum atom has a total of 13 electrons in three orbits. Two are in the first orbit, eight in the second orbit, and three in the third orbit (Figure 1-8). There are positions for 18 electrons in the third orbit. The three electrons in the third orbit are not tightly bound to the nucleus, and the third orbit can momentarily accommodate extra electrons. Aluminum, therefore, is a good conductor of electricity, but *not as good* as copper. If the same size of copper and aluminum wire were used, copper would conduct better. If weight and cost are factors to be considered, aluminum—a lightweight metal—is an excellent choice. With wire of equal length and weight (not size), aluminum will conduct electricity *twice as well* as copper. With equal weights, the aluminum wire will, of course, be much larger.

THE BEHAVIOR OF ELECTRONS

After walking across a rug or sliding out of a car, you sometimes have the strange ability to cause an electric spark when you touch some other object. When you have that ability, you are *electrically charged,* or you carry an electric charge. This same type of electrical charge can be demonstrated by running a plastic comb through dry hair. The comb will attract lightweight objects, such as small scraps of paper or thread.

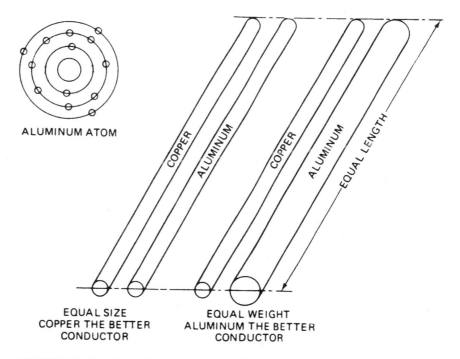

ALUMINUM ATOM

COPPER

ALUMINUM

COPPER

ALUMINUM

EQUAL LENGTH

EQUAL SIZE
COPPER THE BETTER
CONDUCTOR

EQUAL WEIGHT
ALUMINUM THE BETTER
CONDUCTOR

FIGURE 1-8   Relation of weight and size of aluminum and copper conductors.

Static Electricity

The electricity that is obtained by rubbing certain materials together is called *static electricity*. It is called static because electrons are picked up on an item and the electrical charge remains on the item until it touches some other object that does not have a like charge.

As experiments were performed on static electricity, it was found that two types of charges could be obtained, depending on the nature of the materials rubbed together: if two glass rods are rubbed with a silk cloth, they are charged alike and repel each other; if two hard rubber rods are rubbed with wool, they are also charged alike and also repel each other.

CONCLUSION: Like charges repel each other.

If a charged glass rod is brought near a charged rubber rod, the glass rod and rubber rod will be attracted to each other. The charges on the rods are unlike each other, so they attract each other.

CONCLUSION: Unlike charges attract each other.

In order to classify the charges, they were given names. Benjamin N. Franklin (1706–1790), an American scientist, suggested *positive* and *negative* charges. The glass rods have a positive charge; the rubber rods have a negative charge.

Static electricity is electricity at rest.

Dynamic Electricity

Static electricity is not a practical form of electricity. The practical form of electricity that is used to provide power and energy all over the world is dynamic electricity.

Dynamic electricity is electricity in motion.

In an automobile, the battery provides dynamic electricity, as does the alternator (generator). For the power company, water—at high pressure behind dams—moves large alternators that provide dynamic electricity. In some areas where there are no rivers for dams, large steam plants are used turn alternators that generate dynamic electricity.

A new kind of power plant, which uses atomic energy to provide the heat needed for steam turbine operation, has been constructed in many areas of the country. One example is located at Turkey Point, Florida; it produces much of the electricity used in south Florida (Figure 1-9).

FIGURE 1-9 Turkey Point, Florida, power plants.

Another energy source that may be of greater importance in the future, geothermal energy, is presently being investigated. To tap this energy source, a deep hole is drilled into the earth to an area of intense heat. Water is introduced into the hole, where it is converted to superheated steam at the lower level. The steam returns to the earth's surface to operate turbines that rotate the electric alternators that provide electricity.

Regardless of the energy source, dynamic electricity is produced by generators or alternators converting mechanical energy into electrical energy.

VOLTAGE AND CURRENT

A well-developed understanding of the basic terms relating to dynamic electricity and their relationship with each other is necessary in order to troubleshoot air-conditioning and refrigeration electric systems effectively.

Voltage and current are terms associated with dynamic electricity. Voltage and current will be covered together, since it is not possible to obtain one without some of the other.

Voltage is the pressure that causes electrons to move.
Current is the movement of electrons.

Compressor Analogy of Electricity

As is often the case, it is easier to understand a new subject when it is related to some past experience. The electrical phenomena will be related to an air-conditioning compressor system.

In this analogy, it is not necessary to be concerned with temperature changes in the refrigeration system; only the gas movement need be considered. Figure 1-10 shows a compressor with inlet (suction) and outlet (discharge) lines attached. The lines are sealed at the ends so that no gas can enter or exit the system. When the compressor is not running, the system will be in *balance* with an equal number of gas molecules on either side.

A balance is a state in nature that everything is trying to reach.

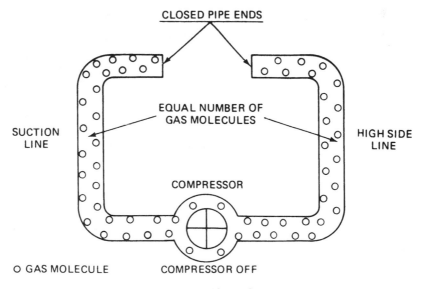

FIGURE 1-10 Balanced system.

Figure 1-11 shows the compressor in operation. Some of the gas molecules are moved by the compressor from the suction line (on the left) to the discharge side (on the right). There is a relative excess of gas molecules in the discharge line (compression), causing a pressure difference between the two lines.

If a small pipe, as shown in Figure 1-12, were connected from the discharge line to the suction line, some of the gas molecules would move through the pipe. There would be continuous movement of gas molecules through the system as long as the compressor is operating. The suction side of the system will continue to be under a vacuum, and the discharge side will continue at high pressure as long as the small tubing offers some restriction (resistance) to the movement (flow) of gas molecules through the system.

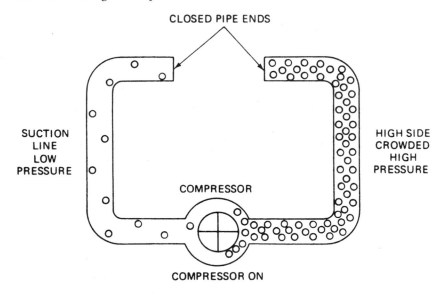

FIGURE 1-11   Unbalanced system.

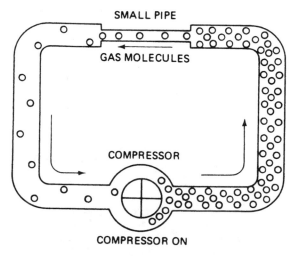

FIGURE 1-12   Molecule movement.

This system should seem reasonable to the student of refrigeration and air-conditioning technology, since it is similar to the basic system most commonly associated with the refrigeration cycle. It is equally reasonable with electricity.

In Figure 1-13, an electric generator has been substituted for the compressor, and the air lines have been replaced with copper wire. The circuit is open, since there is no connection between the wire from the left side of the generator and the wire from the right side. Within the copper wire, there are a large number of copper atoms. Each atom has one free electron (easily moved) in its outer orbit.

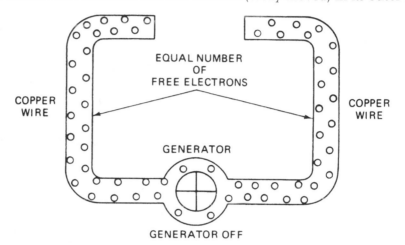

EQUAL NUMBER
OF
FREE ELECTRONS

COPPER
WIRE

COPPER
WIRE

GENERATOR

GENERATOR OFF

FIGURE 1-13 Electrical circuit in balance.

The electrical generator, when running, is capable of moving electrons. This is similar to the compressor's ability to move gas molecules. How the generator moves electrons is covered in later units.

When the generator is not running, there are an equal number of free electrons in the wires on either side of the generator: the system is in balance. When turned on, the generator draws some of the free electrons from the left side and pushes them to the right side, as in Figure 1-14. The movement of electrons from one side to the other creates an *unbalance.* As long as there is an unbalance, there will be a *pressure* called *voltage.* The pressure is measured in volts and is exerted in an effort to try to balance the system. Another term for voltage is *electromotive force,* abbreviated emf.

Since there is an excess of electrons on the right side, this side of the generator is negative (or we could say there is an absence of electrons on the left side). Since the nucleus of the copper atoms has not changed, the protons remain the same on both sides of the generator. The situation now is that there are more protons on the left side than electrons. Thus, the left side of the generator is positive. As long as the generator is operating, the system will remain unbalanced; voltage will exist *between* any point on the right side and any point on the left side.

If a voltage-measuring device (called a *voltmeter*) is connected across the open wire ends, the meter needle will move (Figure 1-15). This indicates the unbalance in electrical pressure that exists *between the two points.* Voltage is always measured between two points in a circuit. (See Unit 3.)

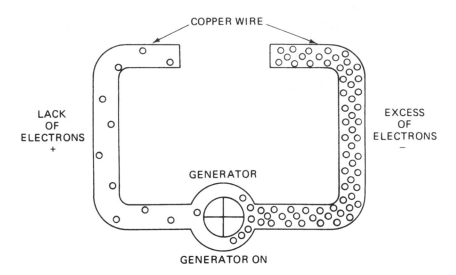

FIGURE 1-14   Unbalanced voltage.

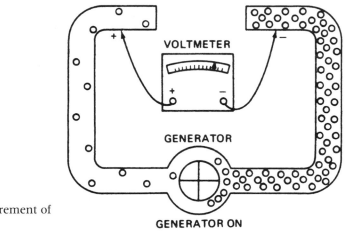

FIGURE 1-15   Measurement of voltage.

WARNING: *Never connect a wire across the output terminals of a generator. The wire will short out the generator. It is used here only as a means of explanation.*

If a small wire is connected between the left side of the generator and the right side, some of the excess electrons will move from the right side through the wire to the left side (Figure 1-16). The generator will continue to move electrons, just as the compressor moves gas. An electric current will flow in the circuit as long as the circuit is complete and the generator is running. As long as there is an unbalance and a circuit, current will flow. Current, or electrons in motion, is measured in amperes, commonly referred to as amps. It takes 6,250,000,000,000,000,000 (6.25×10^{18}) electrons *passing a point in one second to* make 1 ampere. This quantity is called a *coulomb*.

Voltage is caused by an unbalance in electrons. It is *not* the electrons but *the electrons that have moved* that create the unbalance. Similarly, the current in the circuit is the electrons in motion.

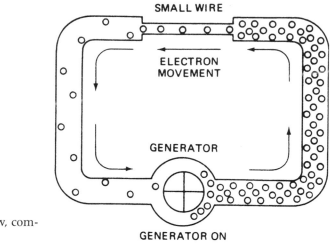

FIGURE 1-16   Electron flow, complete circuit.

SUMMARY

- Whenever there is an imbalance in an electrical system, a pressure will exist trying to balance the system.

- In an electrical system the unbalance, or pressure between the two points in the system, is the electromotive force (emf)—most often called voltage. Remember: voltage is measured between two points.

- If a circuit is provided between the unbalanced points, current (a movement of electrons) will flow through the circuit.

- If a circuit is not connected between the unbalanced points (open circuit), no current will flow.

PRACTICAL EXPERIENCE

Required equipment   A plastic or rubber pocket comb and paper.

Exercise 1
1. Tear a small piece of paper into smaller pieces, say, a half-inch (12–13 mm) square.
2. Pass the comb through your hair several times.
3. Place the comb near the pieces of paper.

Quiz
1. Did the comb attract the pieces of paper?
   (a) If no, before proceeding, determine why not.
   (b) If yes, proceed with question 2.

2. What do you think caused this attraction?

3. Is this exercise an example of static or dynamic electricity?

4. Explain.

Exercise 2

1. On a wall-type air-conditioning unit, locate the power plug.

2. Observe the shape of the power plug.

3. Draw the shape of the power plug on a separate sheet of paper.

4. Locate the nameplate on the air-conditioning unit.

5. Find the voltage rating of the unit on that nameplate.

6. If the voltage rating of the unit is 120 volts (V), the power plug should be shaped like large, standard home appliance plugs such as shown in Figure 1-17.

END VIEW

FIGURE 1-17 A 120-V power plug.

7. Is that the case?

8. If the voltage rating of the unit as indicated on the nameplate were 220 V at less than 15 amperes, the plug would be as shown in Figure 1-18.

END VIEW

FIGURE 1-18 A 220-V power plug.

9. Examine other air-conditioning-unit plugs.

10. Determine the voltage and current limitations of the plugs by their shape.

11. Check the air-conditioner nameplate for voltage and current levels.

12. Visit a hardware or home maintenance store. Check in the electrical department for different types of outlets and plugs with their voltage and current rating.

REVIEW QUESTIONS

1. Static electricity is electricity at rest.   T_____   F_____
2. Dynamic electricity is electricity in motion.   T_____   F_____
3. Like charges repel each other.   T_____   F_____
4. Unlike charges repel each other.   T_____   F_____
5. Voltage is a pressure that causes electrons to move. T_____   F_____
6. Current is the movement of electrons.   T_____   F_____
7. The natural state is for everything to be in a
   balanced condition.   T_____   F_____
8. Current is measured in amperes.   T_____   F_____
9. Electromotive force, emf, is measured in volts.   T_____   F_____
10. With equal size and length of conductors,
    copper is a better conductor than aluminum.   T_____   F_____

Unit 2

Magnetism

OBJECTIVES

After study and review of this unit, you will understand the relationship of
• Natural magnets and artificial magnets.
• Magnetic polarity and magnetic fields.
• Electromagnets and magnetic strength.

Magnets and magnetism are involved in many components of air-conditioning and refrigeration electrical circuits. Compressors are driven by electric motors that function through magnetism. Many electrical controls operate on magnetic principles. A basic understanding of magnetic principles and magnetic circuits will be helpful to all air-conditioning and refrigeration technicians.

NATURAL MAGNETS

Ancient peoples knew of the special properties of certain stones, notably those found in Magnesia, Asia Minor, that attracted small bits of iron. The stones are called *magnets,* after the location in which they were found. Since most ancient peoples had little scientific knowledge, the action of the magnets was attributed to magic. Natural magnets are stone with a high iron-ore content. Because of its special characteristics, this stone was given the name *magnetite.*

Around the year A.D. 1000, the property of natural magnets to point in a fixed direction was discovered. Since these stones were now useful in navigation, they were given a new name, *loadstone* (also spelled lodestone). The name loadstone means "leading stone" and is still used to indicate special iron ore with magnetic characteristics.

ARTIFICIAL MAGNETS

A piece of iron or steel can be magnetized by placing it in contact with a natural magnet or loadstone. Another method of producing an artificial magnet is to wrap a coil of wire around the iron or steel core and then to pass an electrical current through the wire.

If the material being magnetized is hardened steel, it will retain its magnetism for a considerable length of time. If soft iron is used, a strong magnetism will appear during the magnetizing process, but little magnetism will remain after the magnetizing force is removed. Permanent magnets are made from hardened steel and other special alloys because of their characteristics. Soft iron or steel is used as the core of relays and solenoids, where strong magnetism must be present only when current is flowing in the coils.

MAGNETIC POLARITY

If a piece of metal, such as a steel bar, is magnetized, the magnetic effects concentrate at the ends of the bar. These points of magnetic concentration are called the *poles* of the magnet. Away from the ends and around the bar, an invisible force is present, the effects of which can be seen if small bits of iron or steel are placed near the magnet. The invisible force around the magnet is called a *magnetic field.*

The planet earth is a permanent magnet, with one end near the north pole and the other near the south pole (Figure 2-1). If a bar magnet were balanced near the center by a small thread, it would tend to line up with the earth's magnetic field. One end of the bar magnet would point north. It is therefore labeled the north-seeking pole and usually is marked with the letter N. The other end of the bar would be pointing south and is marked S, for south-seeking pole (Figure 2-2).

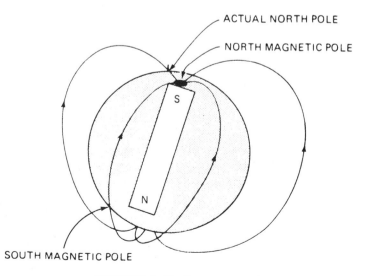

FIGURE 2-1 Earth as a magnet.

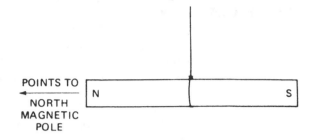

POINTS TO
NORTH
MAGNETIC
POLE

A BAR MAGNET BALANCED AND HUNG
BY A STRING WILL ALIGN ITSELF WITH
THE EARTH'S MAGNETIC FIELD

FIGURE 2-2   Bar magnet compass.

By definition, the magnetic field of a bar magnet is said to emerge from the north pole and to re-enter the magnet at the south pole. Some lines will emerge from the sides of the magnet, but the major concentration is at the poles (Figure 2-3).

If two bar magnets are hung at their balance points and then brought together, the repelling and attracting forces can be demonstrated. In Figure 2-4, two magnets are hung from strings, and the repulsion of two like poles is shown. As the magnetic bars approach each other, the two north poles swing away from each other.

In Figure 2-5, the attraction of two unlike poles is shown. As the two unlike poles approach, they are attracted to each other, as the pull on the strings indicates.

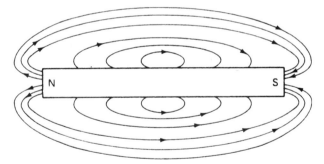

FIGURE 2-3   Magnetic field.

FIGURE 2-4   Like poles repel. (Courtesy of BET Inc.)

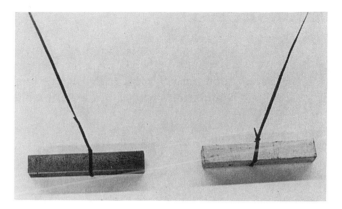

FIGURE 2-5 Unlike poles attract. (Courtesy of BET Inc.)

MAGNETIC MATERIALS

Most materials are not magnetic. Copper, aluminum, brass, wood, paper, and glass are examples of nonmagnetic materials. Magnetic fields will pass through these materials, but the material will not be magnetized. Some materials are magnetic; iron, steel, cobalt, and nickel are examples of magnetic materials. Many alloys of iron, cobalt, and nickel also provide excellent magnetic properties. Some of these alloys contain nonmagnetic materials, such as aluminum, but the final alloy is magnetic and, in most cases, very highly so.

MOLECULAR MAGNETS

In ordinary steel, each molecule of steel is a tiny, permanent magnet. In unmagnetized steel, the molecular magnets are arranged haphazardly throughout the metal (Figure 2-6). Since each molecular magnet is pointing in a different direction, the total magnetic effect is zero. The steel is not a magnet and will not attract magnetic materials.

When steel becomes magnetized, the magnetizing process rotates the molecular structure (magnets) so that most molecules are pointing in the same direction. This arrangement is shown in Figure 2-7.

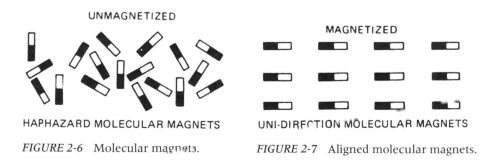

FIGURE 2-6 Molecular magnets.

FIGURE 2-7 Aligned molecular magnets.

MAGNETIC FIELD

Although the magnetic field is invisible, its effect can be demonstrated. If a piece of paper is placed over a bar magnet and the paper is sprinkled with iron filings, a pattern will form. The magnetic force, or field, will arrange the filings in lines running from one end of the bar magnet to the other, as shown in Figure 2-8.

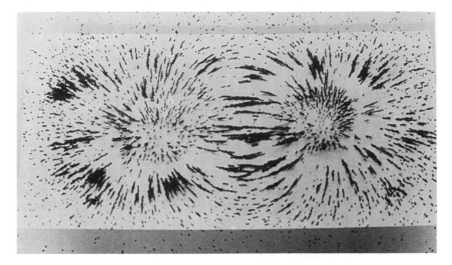

FIGURE 2-8   Iron filings show field around bar magnet. (Courtesy of BET Inc.)

There are no insulators from magnetic fields. Magnetic lines of force can pass through all materials. They will be deflected or bent by other magnetic fields but not stopped or blocked.

The term *antimagnetic* is in most cases a play on words. A wrist watch, for example, may be marked antimagnetic. Since the delicate mechanisms within a watch may be very susceptible to magnetic fields, these small, delicate parts are protected by the frame of the watch, which is made of soft iron (Figure 2-9). Magnetic fields will pass through the iron frame rather than the hard steel parts. The small parts are protected, not because they are antimagnetic, but because the magnetic fields take the easiest path through the soft iron.

ELECTROMAGNETISM

As indicated previously, a magnet may be created by passing an electric current through a coil of wire. This may be explained by one of the basic tenets of electricity: a magnetic field exists about each electron in motion. When electrons move through a wire, the magnetic field set up around the wire is proportional to the amount of current in the wire. This is a basic relation between electricity and magnetism. The magnetic field around a wire carrying current is in the form of concentric circles (Figure 2-10).

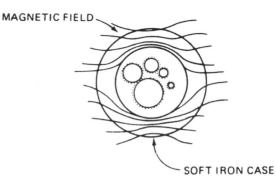

FIGURE 2-9   Magnetic shielding.

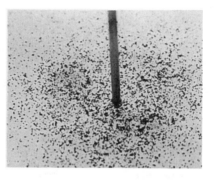

FIGURE 2-10(a)   No current.

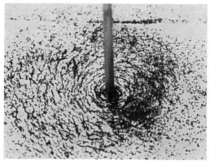

FIGURE 2-10(b)   Current flow.
(Courtesy of BET Inc.).

The method for demonstrating the magnetic field about a wire is shown in Figure 2-11. A wire carrying current is passed through a sheet of paper. Compasses placed on the paper demonstrate the existence of a magnetic field, as well as the direction of the field. The compass needles align themselves with the magnetic field.

LEFT-HAND RULE

If the direction of electron flow in a wire is known, the direction of the magnetic field about the wire can be determined by following the left-hand rule: grasp the current-carrying wire with the left hand with the thumb pointing in the direction of electron flow. The fingers will point in the direction of the magnetic lines of force (Figure 2-12).

FIGURE 2-11(a)   No current, compasses all point north. (Courtesy of BET Inc.)

FIGURE 2-11(b)   Electrons traveling down, compasses line up with field. (Courtesy of BET Inc.)

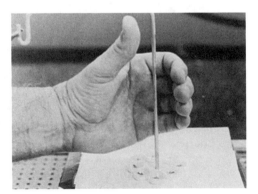

FIGURE 2-12   Left-hand rule. (Courtesy of BET Inc.)

ELECTROMAGNETS

If a current-carrying wire is formed into a loop or coil, the concentric lines of force will all be in the same direction through the center of the loop (Figure 2-13). If loops are placed close together, there will be a further concentration of lines of force. Some lines will merge and go around the combined loops, as shown in Figure 2-14.

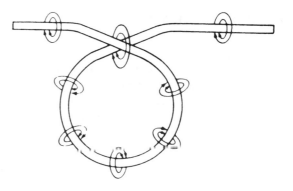

FIGURE 2-13   Field about wire.

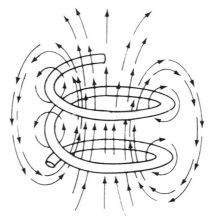

FIGURE 2-14   Field about coil.

If several turns of insulated wire are formed into a coil, lines of force will enter one end of the coil, pass through it, and emerge at the other end. The lines of force will be completed outside of the coil, as shown in Figure 2-15.

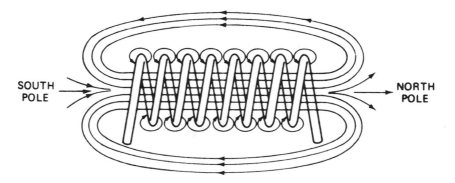

FIGURE 2-15   Concentration of magnetic lines.

MAGNETIC POLARITY

The polarity of an electromagnet may also be determined by using the left-hand rule. The direction of electron current must be known in order to use the rule. Grasp the coil in the left hand with the fingers pointing in the direction of electron flow. The thumb will point in the direction of the lines of force through the center of the coil and to the north pole of the electromagnet (Figure 2-16).

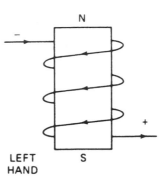

FIGURE 2-16   Left-hand rule.

MAGNETIC STRENGTH

The strength of an electromagnet depends on the size, length, material of the core, number of turns in the coil, and amount of current flowing through it. Other conditions being the same, an electromagnet with a soft iron core will be stronger than one with a core of the steel.

    If the direction of the field is known but the direction of electron flow is not, the rule can be used to determine the current direction. Grasp the wire with the left hand with fingers pointing in the direction of the magnetic lines. The thumb will point in the direction of electron flow.

AMPERE-TURNS

The magnetizing force of a coil is based on the ampere-turns of the coil. The ampere-turns of a coil are equal to the amount of current flowing through the coil multiplied by the number of turns. In Figure 2-17, the ampere-turn rating of different coil-current combinations is given.

1 TURN 1 AMP	2 TURNS 5 AMPS	5 TURNS 2 AMPS	3 TURNS 10 AMPS
TURN x 1 AMP = 1 AT	2 TURNS x 5 AMPS = 10 AT	5 TURNS x 2 AMPS = 10 AT	3 TURNS x 10 AMPS = 30 AT

FIGURE 2-17   Ampere turns.

The magnetizing force of a one-turn coil with 1 ampere of current flowing through it is very weak: 1 ampere-turn. A two-turn coil with 5 amperes flowing through it has a magnetizing force ten times stronger, or 10 ampere-turns. A five-turn coil with 2 amperes of current has the same magnetizing force, 10 ampere-turns. A three-turn coil with 10 amperes of current has the strongest magnetizing force of the group, 30 ampere-turns.

IRON CORE COILS

When a magnetic material is used as the core of an electromagnet, the strength of the magnet is greatly increased. This is very important in the design of components such as electric motors, relays, and solenoids. With an iron core, a stronger magnet is obtained in a coil with the same current flow. This produces a stronger magnet at a lower operating cost.

High-energy, efficient motors are those with more iron in the core and more windings in the coils. The higher initial cost is usually offset in a year or so by savings in operating costs.

A large number of devices that operate on the magnetic principle are found in air-conditioning and refrigeration electrical systems. Specific information on how these devices function will be found in later units of this text.

SUMMARY

- Permanent magnets are made with hardened steel.
- Soft iron is used in temporary magnets.
- Like poles of magnets repel each other.
- Unlike poles of magnets attract each other.
- There are no insulators for magnetism or magnetic lines of force.
- Soft iron is often used to shield items from magnetism by providing a better path for the magnetic lines of force.
- When an electric current flows in a wire, a magnetic field exists around the wire.
- When a current-carrying wire is formed into a loop or coil, the magnetic field is concentrated in the center of the loop.
- When an iron core is used as the core of a coil carrying an electric current, the magnetic field is greatly increased.

PRACTICAL EXPERIENCE

Equipment required   A permanent magnet; copper, aluminum and iron material; and a screwdriver.

Procedure

1. In sequence, place the magnet close to the copper, aluminum, and iron. Describe the action that took place as the magnet was placed close to the suggested materials.

2. Place the magnet at the screwdriver point. Did the magnet and screwdriver attract each other?

3. Stroke the screwdriver with a pole of the magnet from about 2 inches from the tip to the tip.

4. Place the tip of the screwdriver at the piece of iron. Is there an attraction exhibited? The action in steps 3 and 4 demonstrate the ease with which magnetic materials iron and steel may be magnetized.

Conclusion Certain magnetic materials, iron and steel, are easily magnetized when they come into contact with a magnet.

REVIEW QUESTIONS

1. With magnets, like poles (repel, attract)_____ each other.

2. The concentration of a magnetic field in a bar magnet is at the (side, ends)_____ of the bar.

3. Two adjacent wires carrying current in the same direction will (attract, repel)_____ each other.

4. Two adjacent wires carrying current in the opposite direction will (attract, repel)_____ each other.

5. If a wire carrying current is formed into a loop or coil, the magnetic field will (increase, decrease)_____.

6. If a magnetic material, such as soft iron, is used as the core of an electromagnet, the strength of the magnet will (increase, decrease)_____.

7. Soft iron is used as the core of (permanent, temporary)_____ magnets.

8. Hard steel is used as the core of (permanent, temporary)_____ magnets.

9. (Nothing, glass)_____ will isolate a material from magnetic lines of force.

10. Electric motors containing more iron and windings are usually (cheaper, more expensive)_____ to operate.

Ohm's Law and the Electric Circuit

OBJECTIVES

Upon completion and review of this unit, you will be able to
- Recognize resistors as components in electrical/electronic circuits.
- Recognize the schematic symbol for fixed resistors.
- Recognize the schematic symbol for variable resistors.
- Use Ohm's law to determine one of the three values—voltage, current, or resistance—when two of the values are known.

In Units 1 and 2, voltage and current were covered. There is another factor that is of equal importance in the study of electricity and electrical circuits: *resistance*. In the discussion of voltage and current, the following observation was made: when a generator is operating, electrons are pulled from the wire on one side and forced out on the wire on the other side; this creates an unbalance called voltage. It is shown in Figure 3-1. As long as the generator operates, it will keep the unbalance and a voltage will exist between the two wires.

Note in the circuit of Figure 3-1 that no wire is connected between the output wires of the generator. The resistance of air is extremely high and may be considered infinite. With such a high resistance, no current will flow.

If a small wire is connected between the ends of the original wires, a current will flow in the circuit, as shown in Figure 3-2. The amount of current that flows in the circuit depends on the type of material of which the wire is made, the size (diameter) of the wire, and its length. In the circuit in Figure 3-2, the small wire is said to offer resistance to the current flow. Although resistance is offered, current will flow.

The connecting wire in the circuit in Figure 3-3 is the same length and of the same material. However, the cross-sectional area is twice the size of the wire used in Figure 3-2. More current flows in the circuit in Figure 3-3 than in the circuit in Figure 3-2 because resistance is reduced. Actually, *twice* as much current flows when the larger wire is used.

The resistance of a wire depends on four things:

1. Material used to make the wire (available free electrons).
2. Cross-sectional area of the wire (size).
3. Length of the wire.
4. Temperature (normally as temperature increases, resistance increases).

Resistors come in many shapes and sizes. Figure 3-4 shows four resistors: three are different types of wire-wound power resistors, and one is a carbon resistor.

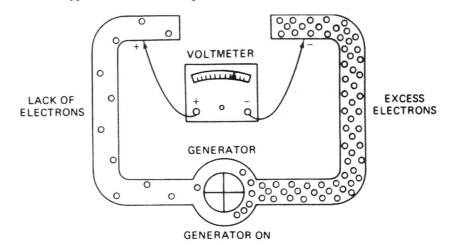

FIGURE 3-1   Generator producing voltage.

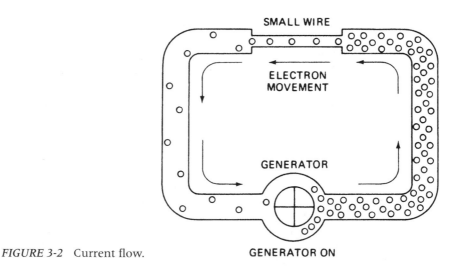

FIGURE 3-2   Current flow.

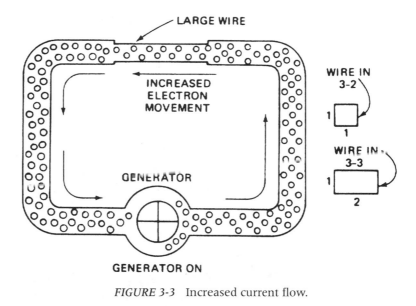

FIGURE 3-3   Increased current flow.

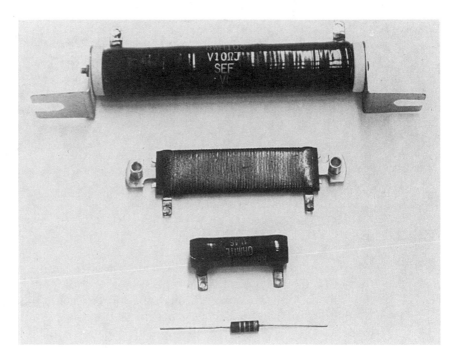

FIGURE 3-4   Resistors. (Courtesy of BET Inc.)

Another resistive element with which you may become familiar is the heat strip from an air-conditioning system (Figure 3-5). The component is made of nichrome wire that has been coiled to decrease the space needed to accommodate the total wire length. The heat strip is a resistive element.

FIGURE 3-5   Heat strip. (Courtesy of BET Inc.)

COMPLETE ELECTRIC CIRCUITS

For current to flow in an electric circuit, the circuit must be complete. Another way of stating this is that current must be able to flow from the source through the load, back to the source, through the source back to the load, and so on.

In Figure 3-6, an electric generator is the energy source, and a lamp is the load. As long as the circuit is complete, current will flow from the generator (source) through the lamp (load) and back through the generator. There is a continuous path for current to flow in. It is a complete electric circuit.

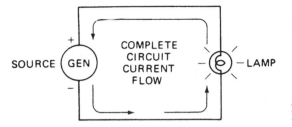

FIGURE 3-6 Complete circuit allows current flow. Lamp is lit.

If a switch is included in the circuit, the same current that flows through the lamp flows through the switch. Two conditions could exist. When the switch is closed, the circuit is complete and current can flow (Figure 3-7). When the switch is in the open position, the circuit is not complete (Figure 3-8). Current cannot flow through the open switch; therefore, current cannot flow through the lamp.

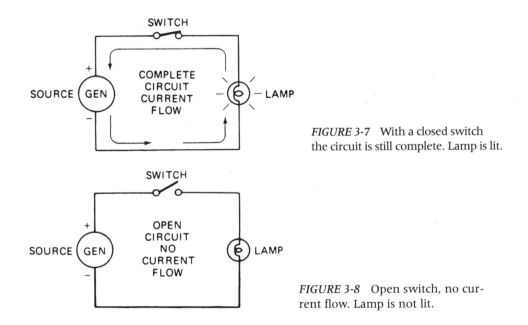

FIGURE 3-7 With a closed switch the circuit is still complete. Lamp is lit.

FIGURE 3-8 Open switch, no current flow. Lamp is not lit.

Figure 3-9 shows two circuits. Note the position of the switches. In both circuits, the switches and the lamp are connected in series. In the circuit in Figure 3-9a, if either switch A-1 or A-2 is open, the circuit is not complete: the lamp will not light. Similarly, in the circuit in Figure 3-9b, if either switch B-1 or B-2 is open, the circuit is not complete. The lamp will not light.

A complete circuit may be made up of the generator, the load, and the connecting wires.

The connecting wires in a circuit are usually considered not to have resistance. Actually, wires do have resistance; but, for practical application, the resistance of wires (copper) is so small that it need not be considered in circuit calculation.

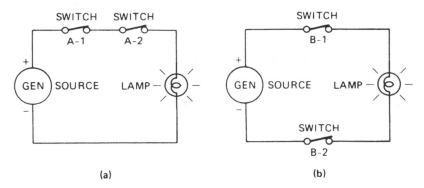

FIGURE 3-9   Switches may be located in different parts of the circuit.

Electrically, both circuits in Figure 3-9 are the same. They are both series circuits. The same current must flow through each component of the circuit.

Figure 3-10 shows a simple control circuit for an air-conditioning compressor contactor. It is a series circuit. The contactor will be energized as long as 24 volts are available from the source and both the *high-pressure cutout* and the *thermostat* select switches are closed. If either switch should open, the circuit to the contactor coil will open, no current can flow in the circuit, and the contactor will de-energize and remove power from the compressor motor.

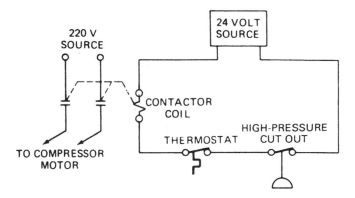

FIGURE 3-10   Low-voltage circuit controlling a high-voltage circuit.

Consider Figure 3-11, a power source connected to a group of lamps. In this circuit, switch S1 is closed. There is a complete circuit from the A terminal of the generator through S1, through the lamp L1, to generator B and through the generator to A. The circuit is complete. Lamp L1 will be lit.

If switch S2 is closed, a complete circuit exists from generator A through S2, through lamp L2, and back through the generator. Lamp L2 will be lit.

When switch S3 is open, a complete path does not exist for current flow in the L3 lamp circuit. Lamp L3 will not be lit.

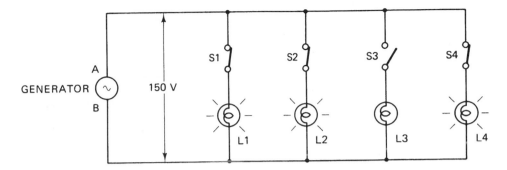

FIGURE 3-11 Parallel lamp circuit with individual switches.

If switch S4 is closed, a complete circuit exists through S4, the lamp L4, and the generator. Lamp L4 will be lit.

If current is to flow in any electric circuit, a complete path must exist from the power source, through the load, and back to the power source.

OHM'S LAW

The relationship of voltage, current, and resistance may be determined by a relatively simple formula known as Ohm's law. It was devised by a German physicist, Georg Simon Ohm, in 1848. He discovered that the current in any circuit is directly proportional to the voltage applied and is inversely proportional to the resistance.

$$I = \frac{E}{R}$$

or

$$R = \frac{E}{I}$$

or

$$E = I \times R$$

From Ohm's law, the following can be derived:

1. If the resistance is kept the same and the voltage is increased, the current will increase.
2. If the voltage is kept the same and the resistance is increased, the current will decrease.

In honor of the discovery of the law, the term for resistance measurement is the *ohm;* it is indicated by the Greek letter omega Ω. The symbol for resistance is the letter R. The symbol for voltage is the letter E (from emf). The symbol for current is the letter I (from current intensity).

By the use of any one of the three forms of Ohm's law, when two factors are known in a circuit, the third factor can always be found.

EXAMPLE 1   (*I* is the unknown)

A 12-ohm resistor has 24 volts across it. How much current is flowing through it (Figure 3-12)?

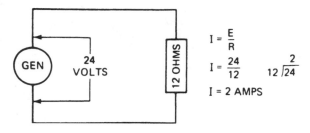

$$I = \frac{E}{R}$$

$$I = \frac{24}{12} \qquad 12\,\overline{)24}$$

$$I = 2 \text{ AMPS}$$

FIGURE 3-12 Using Ohm's law to determine current.

Solution

$$I = \frac{E}{R}$$

$$I = \frac{24}{12}$$

$$I = 2 \text{ amps.}$$

Mathematically, the answer is 2 amperes. If an ammeter were connected in series with the 12-ohm resistor, as in Figure 3-13, the meter would indicate 2 amperes.

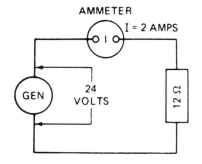

FIGURE 3-13 Connecting ammeter to indicate current.

EXAMPLE 2   (*E* is the unknown)

A resistor of 60 ohms is connected across an electrical power source. An ammeter connected in series with the resistor indicates 2 amperes (Figure 3-14). What is the voltage of the power source?

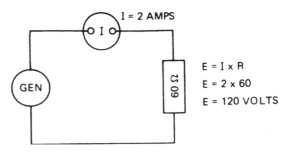

FIGURE 3-14   Using Ohm's law to
determine voltage.

Solution

$$E = I \times R$$

$$E = 2 \times 60$$

$$E = 120 \text{ volts}$$

    Mathematically, the answer is 120 volts. If a voltmeter were connected across
the power source, it would indicate 120 volts (Figure 3-15).

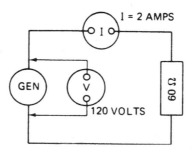

FIGURE 3-15   Measuring supply
voltage.

EXAMPLE 3   (R is the unknown) _____

    A resistor is connected across a 220-volt power source. An ammeter con-
nected in series with the resistor indicates 10 amperes (Figure 3-16). What is the
value of the resistor?

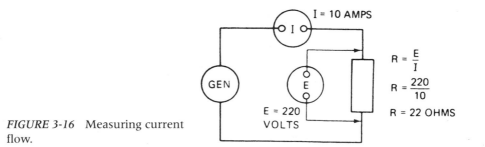

FIGURE 3-16   Measuring current
flow.

Solution

$$R = \frac{E}{I}$$

$$R = \frac{220}{10}$$

$$R = 22 \text{ ohms.}$$

Mathematically, the answer is 22 Ω. If an ohmmeter were connected to the resistive device, it would indicate 22 Ω (Figure 3-17).

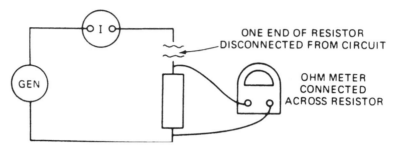

FIGURE 3-17   Using an ohmmeter to measure resistance.

In normal, everyday operations, the air-conditioning and refrigeration technician is not required to solve Ohm's law problems. An understanding of Ohm's law does, however, provide the means to a better understanding of electricity. A good understanding of electricity is important to the technician when troubleshooting equipment.

Open Circuit—Ohm's Law

Ohm's law is true for the whole circuit or any part of a circuit. Consider the circuit of Figure 3-18a. Switch S1 is closed. The resistance of the closed switch could be 0.001 ohm or even lower. The switch resistance is so low compared to the lamp resistance that the switch resistance does not effectively contribute to the total resistance of the circuit (144 ohms versus 144.001 ohms). By Ohm's law, the current in the circuit is

$$I = \frac{E}{R}$$

$$I = \frac{120}{144}$$

$$I = 0.833 \text{ ampere}$$

In Figure 3-18b, switch S1 is open. The resistance of the open switch could be 100,000,000 ohms. The resistance of the lamp is insignificant when compared to the open switch resistance. The supply voltage, 120 volts, will appear across the open switch. By Ohm's law, the current is

$$I = \frac{E}{R}$$

$$I = \frac{120}{100,000,000}$$

$$I = 0.0000000012 \text{ ampere}$$

Lamp L1 will not be lit as there is too little current flowing through it. For all practical purposes, there is no current. (The actual resistance of the open switch would depend on a number of factors. One is air resistance. The resistance of air varies according to moisture content, among other factors.)

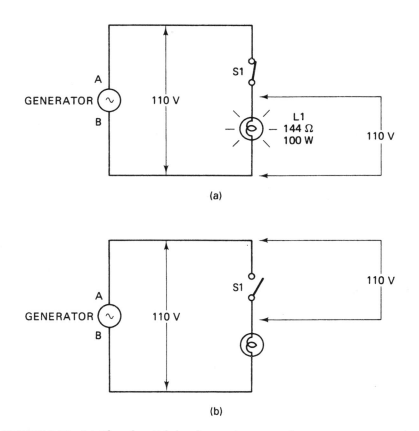

FIGURE 3-18 (a) Closed switch has low resistance; voltage appears across the lamp. (b) Open switch has extremely high resistance; voltage appears across the switch.

SUMMARY

- A complete circuit is necessary if current is to flow.
- Ohm's law is the relationship between current voltage and resistance in an electric circuit.
- If two of the factors in Ohm's law are known, the third can always be found.
- Ohm's law is true for the entire circuit or for any part of the circuit.

PRACTICAL EXPERIENCE

Determine the value of the unknown in the following problems.
1. A 22-ohm resistive heater has 220 volts across it. How much current is flowing through it?
2. A resistor of 15 ohms has 11 amperes flowing through it. How much voltage is across the resistor?
3. The voltage measured across a heat strip is 220 volts. A clamp-on ammeter used on a line feeding this circuit indicates 9 amperes. What is the resistance of the element?

Conclusion

The relationship between current, voltage, and resistance in a component is fixed by a law called Ohm's law. The formula may be stated in three ways:
1. The current flow through a component is equal to the voltage across the component divided by the resistance of the component.

$$I = E/R$$

2. The resistance of the component is equal to the voltage across the component divided by the current flow through the component.

$$R = E/I$$

3. The voltage across the component is equal to the current flow through the component multiplied by the resistance of the component.

$$E = I \times R$$

REVIEW QUESTIONS

1. The resistance of a piece of wire depends on four things. What are they?
2. In an electric circuit, if the resistance is kept the same and the voltage is increased, the current will (increase, decrease)_____.
3. In an electric circuit, if the voltage is kept the same and the resistance is increased, the current will (increase, decrease)_____.

4. If a 6-ohm resistor has 20 amperes flowing through it, the voltage across the resistor will be_____ volts.

5. If 24 volts is measured across an 8-ohm resistor, how much current must be flowing through it?

6. An ammeter connected in series with a resistor indicates 11 amperes. A voltmeter connected across the resistor indicates 220 volts. What is the value of the resistor?

7. What voltage is needed to cause 4 amperes to flow through a 110-ohm resistor?

8. If the voltage across a 15-ohm resistor is measured at 90 volts, how much current is flowing through the resistor?

9. If the voltage applied to a resistor is doubled, will the current double or decrease to one-half the original value?

10. What size resistor will cause a current of 3 amperes to flow when 120 volts is applied to it?

Unit **4**

Series Circuits

OBJECTIVES

Upon completion and review of this unit, you will be able to
- Recognize a complete circuit.
- Recognize a series circuit by component connections.
- Solve series circuit problems.

In Unit 3, Ohm's law and its application to simple electric circuits were discussed. In this unit, the discussion will be expanded to cover Ohm's law as it relates to series circuits.

An example of a simple series circuit is the relay contacts of an air-conditioning compressor and the related compressor motor. (See Figure 4-1.) The contacts are in series with the motor. All the current that flows through the compressor must flow through the contacts. When the contacts are open, the circuit is open; no current flows through the compressor motor.

A more complete presentation of series circuits follows; the relationship between voltage, current, and resistance in a series circuit is covered here.

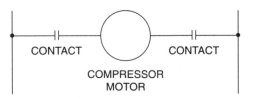

FIGURE 4-1   Contacts in series with compressor motor.

WIRE RESISTORS

A common material used in the construction of wire-wound resistors is nichrome. Nichrome wire has a relatively high resistance for short lengths of wire. For example, 100 feet of nichrome wire could have a resistance of about 100 ohms. A practical, usable resistor can be made out of the nichrome wire by winding it on a core of ceramic with terminals at either end (Figure 4-2).

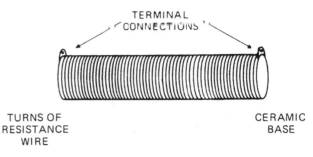

FIGURE 4-2 Wire-wound resistor.

TERMINAL CONNECTIONS

TURNS OF RESISTANCE WIRE

CERAMIC BASE

The circuit in Figure 4-3 consists of a source (a generator), connecting wires, and a load (a 100-ohm wire-wound resistor). According to Ohm's law, if the generator were producing 100 volts, then 1 ampere of current would flow in the resistor. A voltmeter connected across the resistor would read 100 volts.

$$I = \frac{E}{R}$$

$$I = \frac{100}{100}$$

$$I = 1 \text{ ampere}$$

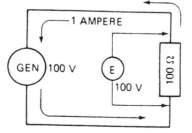

FIGURE 4-3 Single-resistor circuit.

If the resistor were broken at the exact center (Figure 4-4), there would be two resistors of 50 ohms each (two 50-foot lengths of the nichrome wire). The two broken ends of wire could be clamped together to remake the 100-ohm resistor (Figure 4-5). The two 50-ohm resistors, tied in series, make the 100-ohm resistor: 50 + 50 = 100.

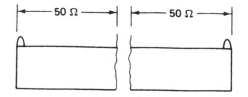

FIGURE 4-4   100 Ω resistor broken
at exact center.

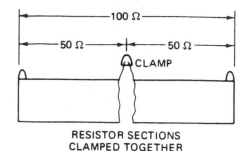

RESISTOR SECTIONS
CLAMPED TOGETHER

FIGURE 4-5   Two 50-ohm resistors
joined together.

The two reconnected 50-ohm resistor sections can be connected into the original circuit (Figure 4-6). This circuit is a series circuit. The same current, 1 ampere, flows through each of the 50-ohm resistor sections. Ohm's law may be used to calculate the voltage across R_1, the first resistor section.

$$E = I \times R_1$$

$$E = 1 \times 50$$

$$E = 50 \text{ volts}$$

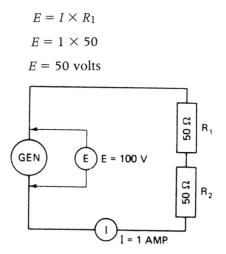

FIGURE 4-6   Two-resistor series
circuit.

Ohm's law is also used to calculate the voltage across R_2, the second resistor section.

$$E = I \times R_2$$

$$E = 1 \times 50$$

$$E = 50 \text{ volts}$$

Note that the sum of the voltages across R_1 and R_2 equals the supplied voltage, 50 V + 50 V = 100 V.

$$E_T = E_{R1} + E_{R2} + \ldots$$

Also note that the total resistance, 100 ohms, is the sum of the individual resistors: 50 Ω + 50 Ω = 100 Ω.

$$R_t = R_1 + R_2 + \ldots$$

We can now state the laws of series circuits.

LAWS OF SERIES CIRCUITS

1. The same current flows through each component of a series circuit.
2. The total voltage in a series circuit is equal to the sum of the voltages across the individual components.
3. The total resistance of a series circuit is equal to the sum of the resistances of the individual components.

Example 1

A circuit consists of a 25-ohm resistor and a 50-ohm resistor connected in series across a 150-volt generator (Figure 4-7). Determine the current flow in the circuit and the voltage across each resistor.

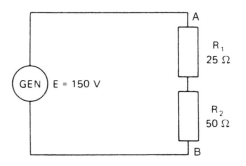

FIGURE 4-7 Circuits of 25 ohms and 50 ohms in series.

Solution   The total resistance is equal to the sum of the individual resistances.

$$R_t = R_1 + R_2$$
$$R_t = 25 + 50$$
$$R_t = 75 \text{ ohms}$$

Use Ohm's law on the whole circuit:

$$I = \frac{E}{R}$$
$$I = \frac{150 \ V}{75}$$
$$I = 2 \text{ amperes}$$

Since the same current flows through each component, the voltage across each component may be determined by using Ohm's law:

$$ER_1 = I \times R \qquad\qquad ER_2 = I \times R$$

$$ER_1 = 2 \times 25 \qquad\qquad ER_2 = 2 \times 50$$

$$ER_1 = 50 \text{ volts} \qquad\qquad ER_2 = 100 \text{ volts}$$

The supply voltage is equal to the sum of the voltages across the individual components.

$$E_T = E_{R1} + E_{R2}$$

$$E_T = 50 + 100$$

$$E_T = 150$$

Example 2

Determine the value of each resistor, the voltage across each resistor, and the total resistance of the circuit shown in Figure 4-8. The generator supplies 240 volts; the ammeter indicates 4 amperes.

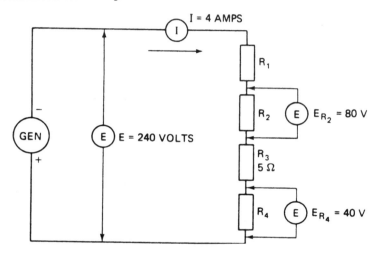

FIGURE 4-8   Using Ohms law in series circuits.

Solution   Solve for R_T, using Ohm's law for the whole circuit:

$$R_T = \frac{E_T}{I}$$

$$R_T = \frac{240}{4}$$

$$R_T = 60 \text{ ohms}$$

Solve for R_2:

$$R_2 = \frac{E_{R2}}{I}$$

$$R_2 = \frac{80}{4}$$

$$R_2 = 20 \text{ ohms}$$

Solve for E_{R3}:

$$E_{R3} = I \times R_3$$

$$E_{R3} = 4 \times 5$$

$$E_{R3} = 20 \text{ volts}$$

Solve for R_4:

$$R_4 = \frac{E_{R4}}{I}$$

$$R_4 = \frac{40}{4}$$

$$R_4 = 10 \text{ ohms}$$

Solve for R_1, using the total resistance law:

$$R_T = R_1 + R_2 + R_3 + R_4$$

$$60 = R_1 + 20 + 5 + 10$$

$$R_1 = 60 - 35$$

$$R_1 = 25 \text{ ohms}$$

Check the solution, using total voltage:

$$R_1 = 25 \text{ ohms}$$

$$E_{R1} = I \times R$$

$$E_{R1} = 4 \times 25$$

$$E_{R1} = 100 \text{ V}$$

$$E_T = E_{R1} + E_{R2} + E_{R3} + E_{R4}$$

$$240 = 100 + 80 + 20 + 40$$

$$240 = 240$$

The solution checks.

The refrigeration and air-conditioning technician will not often have to solve electrical problems mathematically. When the technician is required to solve these problems, it is important to know the laws of series circuits.

SUMMARY

• In a series circuit, the sum of the voltage across the components is equal to the applied voltage.

- The same current flows through each component in a series circuit.
- The total resistance of a series circuit is equal to the sum of the resistances of the individual components.

PRACTICAL EXPERIENCE

Required equipment   Volt-ohm milliammeter (VOM) and resistors with values within the same range (1 to 100), (100 to 1000), (1000 to 10,000), and so on. (At least three resistors are required.)

Procedure

1. Select ohmmeter service on the VOM.
2. Check the zeroing of the meter by shorting the leads together and making adjustment.
3. Measure and record the value of each of the resistors (change the meter scale and rezero the meter as necessary).

      Resistor 1_____

      Resistor 2_____

      Resistor 3_____

      Resistor 4_____

      Resistor 5_____

4. Connect one end of resistor 1 (R_1) to one end of resistor 2 (R_2), as in Figure 4-7.
5. Connect the ohmmeter to measure the resistance of the two resistors in series. From A to B is _____ohms.
6. Add the value of resistor one and resistor two as recorded in step 2.
7. Is the value determined in step 6 equal to the value measured in step 5?
8. Repeat steps 5, 6, and 7 with different resistors.
9. Three or four resistors may be connected in series. The total resistance should be the sum of the individual resistances.

Conclusions

1. The total resistance of a series circuit is equal to the sum of the individual resistors.
2. The sum of the voltages across the components connected in series is equal to the applied voltage.
3. The same current flows through each component in a series circuit.

REVIEW QUESTIONS

Draw a sketch of each circuit before attempting solutions.
1. A circuit consists of four 20-ohm resistors connected in series. The total resistance of the circuit is _____ ohms.

2. A circuit consists of an 8-ohm resistor and a 4-ohm resistor connected in series. The supply voltage to the circuit is 24 volts. How much current will flow in the circuit? How much voltage will be measured across each resistor?

3. A circuit consists of a 10-ohm resistor and a 2-ohm resistor connected to a 24-volt power source. The 10-ohm resistor has 20 volts across it. How much current is flowing in the 2-ohm resistor?

4. A circuit consists of three resistors connected in series to a power source. Each resistor has 6 volts across it. What is the voltage of the power source?

5. A series circuit consists of two resistors connected to a power source of 24 volts. The first resistor has 16 volts across it; the second is a 4-ohm resistor. What is the total resistance of the circuit?

6. The (same, different) _____ current flows through each resistor in a series circuit.

7. Two 10-ohm resistors are connected in a series circuit. The voltage across the circuit is 40 volts. The current through each resistor is _____ amps.

8. Four 8-ohm resistors are connected in a series circuit. The current through one of the resistors is 2 amperes. The voltage supplied to the circuit is _____ volts.

9. A current flow of 3 amperes is flowing in a two-resistor series circuit. The supply voltage is 24 volts. One of the resistors has a value of 6 ohms. The other resistor is _____ ohms.

10. A three-resistor series circuit is made up of 6-ohm resistors. A voltmeter placed across one of the resistors reads 9 volts. The source voltage is _____ volts.

Unit 5

Parallel and Series Parallel Circuits

OBJECTIVES

Upon completion and review of this unit, you will understand that
- Parallel circuits are circuits where the total line current splits into two or more paths before returning to the current source.
- Series parallel circuits are circuits where some of the components of the circuit are connected in series and other components of the circuit are connected in parallel.
- Ohm's law is applicable to the whole circuit or the individual components of the circuit in both the parallel or series parallel circuit.

An example of a parallel circuit is in the connection of the compressor and fan in many air-conditioning condenser units. The fan motor is connected directly across the compressor motor. Whenever the compressor motor is operating, the fan will be operating. (See Figure 5-1.)

Both motors receive the *same* voltage. This is one of the conditions describing a parallel circuit. The *same* voltage appears across components connected in parallel.

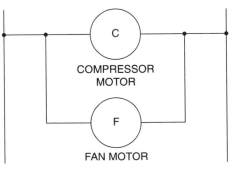

FIGURE 5-1 Compressor motor and fan motor in parallel.

RESISTIVE PARALLEL COMPONENTS

Consider a paper 2 inches (50.8 mm) wide and 4 inches (101.6 mm) long. A very thin layer of carbon material is deposited on this paper forming a resistor, as shown in Figure 5-2. The path for current, then, is 2 inches (50.8 mm) wide by 0.01 inch (0.254 mm) high.

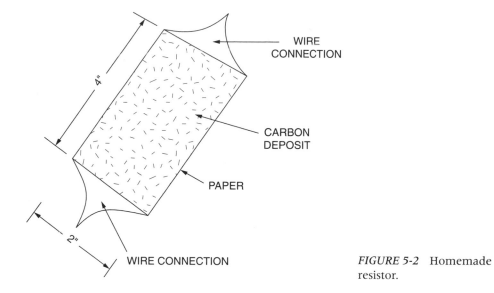

FIGURE 5-2   Homemade resistor.

A voltage of 4 volts is impressed across the resistor and a current of 2 amperes is seen to be flowing through it, as shown in Figure 5-3.

By applying Ohm's law the resistance of the homemade resistor may be calculated:

$$R = V/I \qquad R = 4/2 \qquad R = 2\Omega$$

Next, a very sharp instrument is used to cut our homemade resistor in half lengthwise without removing any of the carbon deposit (Figure 5-4). Since none of the carbon deposit was removed the total circuit was not changed electrically. With 4 volts impressed, a current of 2 amperes flows. The current path is still a total of 2.0 inches (50.8 mm) wide by 0.01 inches (0.254 mm) high.

Since there are now two paths for current to flow in, the current will split. Since the two paths are of the same material and of equal dimensions, the current will split equally. One ampere will flow through the resistor on the left and one ampere will flow through the resistor on the right.

The value of each resistor may be found using Ohm's law. The battery is connected directly across the resistors. Four volts is impressed on each resistor. The current flow through each resistor is 1 ampere. According to Ohm's law, each resistor is

$$R = E/I \qquad R = 4/1 \qquad R = 4\Omega$$

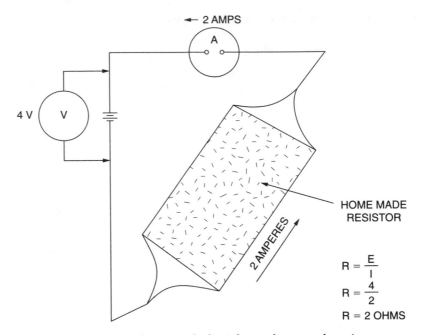

FIGURE 5-3   Application of Ohm's law to homemade resistor.

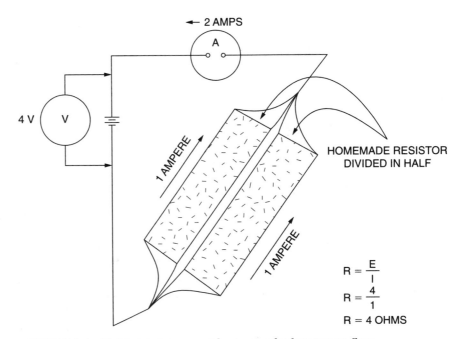

FIGURE 5-4   Divided resistor provides two paths for current flow.

Note that each resistor is 4 Ω while the total resistance of the circuit is only 2 Ω.

The two resistors form what is called a parallel circuit. A parallel circuit, then, is a circuit where there is more than one path for current to flow.

Some other interesting facts may be determined using that same parallel circuit:

1. The same voltage is impressed across each component in a parallel circuit. Many components in circuits may have equal amounts of voltage impressed across them, though they may not be in parallel. If components are in parallel, voltage must be the *same*.

2. The total current is the sum of the currents in the individual branches of the parallel circuit.

3. The effective resistance is smaller than the smallest resistor in a parallel circuit.

PARALLEL CIRCUITS

Parallel circuits are found when the same source voltage is required across two or more components. A typical parallel circuit is shown in Figure 5-5. If the generator is producing 120 volts at its output terminals, there will be 120 volts across each lamp. Current will flow through each lamp. If each lamp has a resistance of 60 ohms, the current will be 2 amperes in each, according to Ohm's law:

$$\frac{120 \text{ volts}}{60 \text{ ohms}} = 2 \text{ amperes}$$

The current for both lamps comes through the generator, the power source. There will be 4 amperes flowing through the generator (2 + 2 = 4).

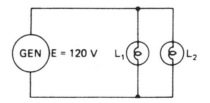

FIGURE 5-5 Parallel circuit.

Current flow in a parallel circuit is similar to water flow in a pipe system. In the water-pipe junction shown in Figure 5-6, 2 gallons of water per minute are flowing in pipe A, entering from the left. In pipe B, 5 gallons of water per minute are flowing, entering from the left. There must be 7 gallons of water per minute flowing out of pipe C. The water cannot disappear. The sum of the water flows into the junction must equal the water flow out of the junction.

In an electric circuit, the sum of the currents entering a junction must equal the current leaving the junction. In Figure 5-7 the lower junction of Figure 5-5 is shown. If 2 amperes are flowing to junction A from lamp L_1, and 2 amperes are flowing to it from lamp L_2, there must be 4 amperes leaving junction A.

Sometimes it is necessary to use Ohm's law to find a branch current value. (A branch is one of the paths for current flow in a parallel circuit.) The circuit in Figure 5-8 is a portion of a total circuit. Ammeter 1 indicates 3 amperes, and ammeter 2, in the lamp circuit, indicates 2 amperes. Since 3 amperes leaves junction B and only 2 amperes come to junction B through the lamp, there must be 1 ampere flowing through resistor R_1 to the junction B.

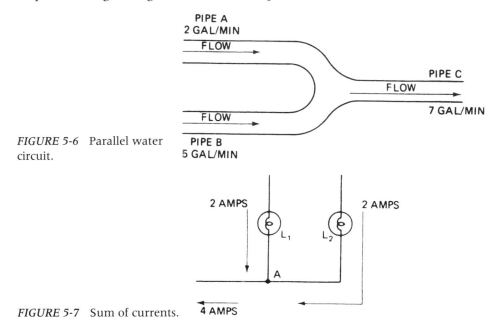

FIGURE 5-6   Parallel water circuit.

FIGURE 5-7   Sum of currents.

The total circuit of Figure 5-8 could contain another branch, as in Figure 5-9. The current entering junction A, coming from junction B, is 3 amperes. The current leaving junction A, going to the generator, is 6 amperes. Where did the other 3 amperes come from? There must be 3 amperes flowing into junction A through R_2. Given that R_2 is an 8-ohm resistor, other unknown values of the circuit may be determined. To determine the voltage across R_2, use Ohm's law for E:

$$E = I \times R$$

$$E = 3 \times 8$$

$$E = 24 \text{ V}$$

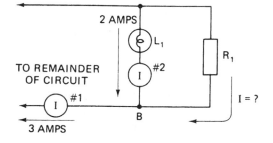

FIGURE 5-8   Determining branch current.

The voltage across each component in parallel is the same. The voltage across R_2 is 24 volts, the voltage across L_1 is 24 volts, and the voltage across R_1 is 24 volts. To determine the resistance of lamp L_1, use Ohm's law for R:

$$R_{L1} = \frac{E}{I}$$

$$R_{L1} = \frac{24}{2}$$

$$R_{L1} = 12 \text{ ohms}$$

To determine the resistance of R_1,

$$R_1 = \frac{E}{I}$$

$$R_1 = \frac{24}{1}$$

$$R_1 = 24 \text{ ohms}$$

To determine the resistance of the circuit, the total current and voltage must be used. The total current as indicated by ammeter 2 (Figure 5-9) is 6 amperes. The total voltage is 24 volts.

$$R_T = \frac{E}{I}$$

$$R_T = \frac{24}{6}$$

$$R_T = 4 \text{ ohms}$$

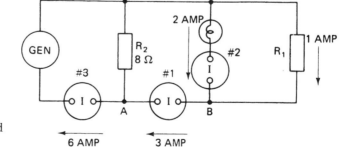

FIGURE 5-9 Branch and line currents.

The total resistance of the circuit is 4 ohms, which is less than the resistance of any of the three components: R_1 (24 ohms), L_1 (12 ohms), or R_2 (8 ohms).

The total resistance of a parallel circuit is less than the smallest branch resistance.

There are other interesting relationships in parallel circuits. The circuit in Figure 5-10 contains a generator producing 24 volts, wires from the generator to a plastic box, and an ammeter indicating 6 amperes. There is no indication as to what is in the box. Solve for R_T of the circuit in the box:

$$R_T = \frac{E}{I}$$

$$R_T = \frac{24}{6}$$

$$R_T = 4 \text{ ohms}$$

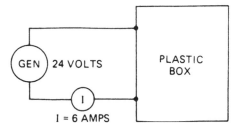

FIGURE 5-10   Black-box analogy.

When the box is opened, a 12-ohm resistor is found connected in parallel with a 6-ohm resistor (Figure 5-11). The resistance of the parallel combination could also be found using the product-over-the-sum method, as in the formula

$$R_T = \frac{R_1 \times R_2}{R_1 + R_2} \quad \frac{\textit{product of } R_1, R_2}{\textit{sum of } R_1, R_2}$$

For the parallel combination in Figure 5-11,

$$R_T = \frac{12 \times 6}{12 + 6}$$

$$R_T = \frac{72}{18}$$

$$R_T = 4 \text{ ohms}$$

This is the same total resistance value that was calculated using Ohm's law.

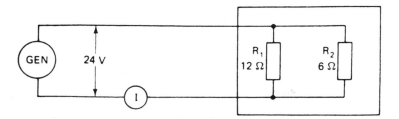

FIGURE 5-11   Inside the black box.

PRODUCT OVER THE SUM

If more than two resistors are connected in parallel, the product-over-the-sum method may be used on two resistors at a time. An example of this procedure is given using Figure 5-12. What is the total resistance of the circuit? It has been shown that the combination of R_1 and R_2 is equal to a 4-ohm resistor. Figure 5-13 shows the equivalent circuit after resistors R_1 and R_2 are combined. The product-over-the-sum method may now be used to determine R_T of the total circuit.

$$R_T = \frac{R_x \times R_3}{R_x + R_3}$$

$$R_T = \frac{4 \times 4}{4 + 4}$$

$$R_T = \frac{16}{8}$$

$$R_T = 2 \text{ ohms}$$

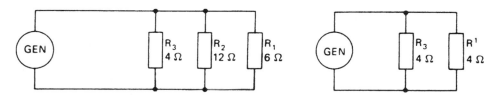

FIGURE 5-12   Three parallel resistors.

FIGURE 5-13   Equivalent circuit.

EQUAL PARALLEL RESISTORS

The R_T of two resistors of equal value connected in parallel is thus equal to the value of one resistor divided by the number of equal resistors. If three 9-ohm resistors are connected in parallel, the R_T will be equal to 9 divided by 3 (Figure 5-14).

There are other methods of solving for the total resistance of parallel circuits that will not be covered in this text. The other methods require higher mathematics. At this time, however, you should understand the following:

1. The same voltage appears across each component of a parallel circuit.

2. The total current in a parallel circuit is equal to the sum of the currents in the branches.

3. The total resistance (R_T) of a parallel circuit is always smaller than the smallest resistor connected in the parallel combination.

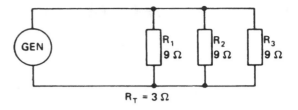

FIGURE 5-14 Equal resistors in parallel.

SERIES PARALLEL CIRCUITS

An example of a series parallel circuit can be made using contactor contacts, a compressor motor, and a condenser fan motor. Observe Figure 5-15. The contactor contacts A and B are in series with the remainder of the circuit, the compressor motor, and the condenser fan motor; the motors are in parallel.

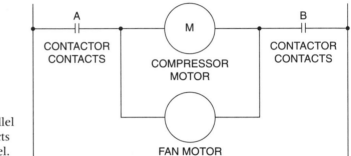

FIGURE 5-15 Series parallel circuit: compressor contacts in series, motors in parallel.

All of the circuit's current must flow through the contactor contacts. The current then splits, with the major portion going to the compressor, and a small portion going to the condenser fan.

Ohm's law may be applied to resistive components connected in series parallel circuits, however, the relationship of the components must be considered when doing so.

A series parallel circuit is made up of a combination of components. Some are connected in series, and others are connected in parallel. When one is working with series parallel circuits, it is necessary to use the laws of series circuits on the series elements and the laws of parallel circuits on the parallel elements.

The circuit in Figure 5-16 is a series parallel circuit. The 10-ohm resistor R_3 is connected in series with the parallel combination R_1 and R_2. To solve for the current flow in the circuit, the resistors must be combined. Resistors R_1 and R_2 are in parallel.

$$R_T = \frac{R_1 \times R_2}{R_1 + R_2}$$

$$R_T = \frac{15 \times 30}{15 + 30}$$

$$R_T = \frac{450}{45}$$

$$R_T = 10 \text{ ohms}$$

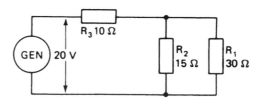

FIGURE 5-16 Series parallel circuit.

The equivalent circuit is a series circuit (Figure 5-17).

$$R_T = R_1 + R_2$$

$$R_T = 10 + 10$$

$$R_T = 20 \text{ ohms}$$

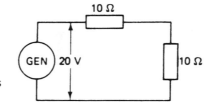

FIGURE 5-17 Equivalent series
circuit.

The generator is producing 20 volts. According to Ohm's law, the current through
the generator is

$$I = \frac{E}{R}$$

$$I = \frac{20}{20}$$

$$I = 1 \text{ ampere}$$

Going back to the original circuit (Figure 5-16), the current through the series
10-ohm resistor R_3 is 1 ampere, and there are 10 volts across it. With 10 volts across
R_3, there are 10 volts left (from the 20 volts supplied) to appear across R_1 and R_2. Since
the *same* voltage appears across resistors in parallel, the current flow through R_1 is

$$I = \frac{E}{R}$$

$$I = \frac{10}{30}$$

$$I = \frac{1}{3} \quad \text{or} \quad 0.333+ \text{ ampere}$$

The current flow through R_2 is

$$I = \frac{E}{R}$$

$$I = \frac{10}{15}$$

$$I = \frac{2}{3} \quad \text{or} \quad 0.666+ \text{ ampere}$$

The total current entering the junction (1 ampere) is equal to the total current in the branches.

$$\frac{1}{3} \quad \text{ampere} \; + \; \frac{2}{3} \quad \text{ampere} \; = \; 1 \quad \text{ampere}$$

$$0.333+ \text{ ampere plus } 0.666+ \text{ ampere} \; = \; 0.999+ \quad \text{or} \quad 1 \quad \text{ampere}$$

SOLVING FOR UNKNOWN RESISTOR

There are times when the information known about a circuit is other than the resistance values. For example, the circuit in Figure 5-16 might be encountered, but the known information could be that of Figure 5-18: the generator is producing 20 volts, the value of resistor R_1 is 30 ohms, the value of resistor R_2 is 20 ohms, and a voltmeter across R_1 indicates 10 volts. The unknown may be found by using the laws of series circuits, parallel circuits, and Ohm's law.

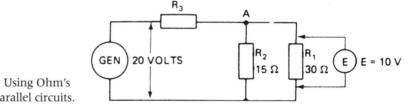

FIGURE 5-18   Using Ohm's law in series parallel circuits.

The current flow through R_1 could be found using Ohm's law. It has already been shown that with 10 volts across 30 ohms the current is $\frac{1}{3}$ ampere. According to the laws of parallel circuits, the voltage across R_2 is the same as across R_1, 10 volts. The current through R_1 can be determined: 10 volts across 15 ohms provides $\frac{2}{3}$ ampere. The current entering junction point A is equal to the sum of the currents leaving it. Therefore, 1 ampere ($\frac{2}{3}$ ampere plus $\frac{1}{3}$ ampere) enters junction point A.

This 1 ampere is the current flowing through the unknown series resistor R_3. Since the supply voltage is 20 volts and there are 10 volts across the parallel resistors R_1 and R_2, there must be 10 volts across R_3 (the supply of 20 volts minus the 10 volts across the parallel branch leaves 10 volts).

According to Ohm's law,

$$R_3 = \frac{E}{I}$$

$$R_3 = \frac{10}{1}$$

$$R_3 = 10 \text{ ohms}$$

SWITCHES IN SERIES PARALLEL CIRCUITS

In air-conditioning and refrigeration electrical systems, switches are used to control the operation of devices such as compressor motors, fan motors, heaters, relays, and indicator lights. These switches are connected in series, parallel, or series parallel as required to accomplish the desired circuit action. Some examples of switch circuit combinations are shown in Figures 5-19a through 5-19f.

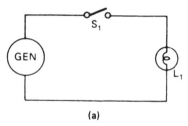

(a)

FIGURE 5-19(a)   Switch S_1 must be closed if lamp L_1 is to light.

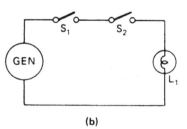

(b)

FIGURE 5-19(b)   Switches S_1 and S_2 must be closed if lamp L_1 is to light.

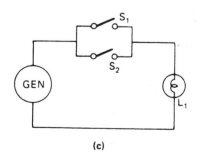

(c)

FIGURE 5-19(c)   Either switch S_1 or S_2 must be closed if lamp L_1 is to light.

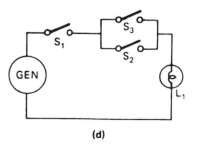

(d)

FIGURE 5-19(d)   Switches S_1 and S_2 or S_3 must be closed if lamp L_1 is to light.

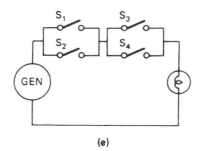

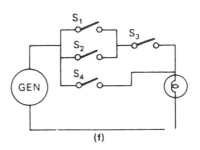

(e)

(f)

FIGURE 5-19(e)   Switches S_1 or S_2 and S_3 or switch S_4 must be closed if lamp L_1 is to light.

FIGURE 5-19(f)   Switches S_1 or S_2 with switch S_3 will cause lamp L_1 to light, or switch S_4 alone will light lamp L_1.

The lamps shown in the preceding circuit diagrams could easily have been compressor motors or fan motors. The switches could have been circuit breakers, temperature controls, or pressure controls. The important point to remember is how switches control various electric alternating current (ac) circuits.

SNEAK CIRCUITS

Whenever switches are connected in series parallel, the possibility of sneak circuits exists. A sneak circuit is one that provides for the operation of a component at a time when operation is not wanted. Sneak circuits are not always obvious at the time a circuit is designed. Some sneak circuits only cause operation under odd conditions and are therefore difficult to discover.

Consider the circuit of Figure 5-20. The path through the switch is normally considered to start from the left top and proceed through the switches to the lamp. Different paths and switches might be considered, but most people would follow a path from left to right. The circuit designer no doubt drew the circuit while considering paths from left to right.

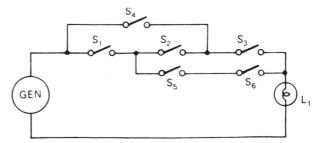

FIGURE 5-20   Switch circuit.

A sneak circuit might exist in a path through S_2 from right to left. If switches S_4, S_2, S_5, and S_6 are closed, there is a complete path for current flow, as shown in Figure 5-21. The circuit path may be an unwanted path causing lamp L_1 to be on during a period when it should not be on. When the sneak circuit is discovered, a redesign is required.

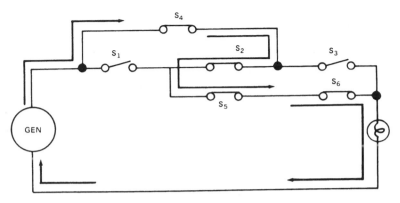

FIGURE 5-21 Sneak circuit.

Most sneak circuits would not be as obvious as the one shown in this example. In some cases, a considerable amount of circuit investigation is necessary before the problem can be discovered. When a sneak circuit is causing a problem, the technician investigating the problem normally assumes that something is malfunctioning, but, in fact, every component is operating correctly.

Sneak circuits are normally in-plant manufacturing problems, not field problems. However, they do sometimes occur in the field and technicians must be aware of them.

SUMMARY

- In parallel resistive circuits, the total current is the sum of the currents in the branches.
- The same voltage appears across each component in parallel.
- The total resistance may be found by using the product-over-the-sum method.
- In series parallel circuits, the laws of series circuits apply to the portions of the circuit connected in series.
- In series parallel circuits, the laws of parallel circuits apply to the portion of the circuit connected in parallel.
- Ohm's law may be applied to the whole circuit or any part of the circuit.

PRACTICAL EXPERIENCE

Required equipment An ohmmeter, a 120-V air conditioner (cover removed), and a wiring diagram of the system.

Procedure
1. Examine the plug attached to the line cord of the air-conditioning (A/C) unit.
2. Determine which plug terminal is the hot terminal and which is the neutral terminal.

3. Follow the neutral terminal to a connection within the A/C unit. (Use the ohmmeter to make this check.)

4. Follow the hot line from the plug to the first connection within the A/C unit.

5. Is there a switch in the hot line controlling the complete A/C unit?

6. Would you consider this switch in series with the rest of the unit?

7. Refer to the wiring diagram of the system.

8. Determine how the thermostat contacts are connected with reference to the compressor motor.

9. Are the contacts in series or parallel?

10. How would you consider the overall connection of the compressor circuit with the fan circuit?

Conclusions

1. Ohm's law is true for the whole circuit or any part of a circuit.

2. In a series parallel circuit, the laws of series circuits are applied to the component connected in series, and the laws of parallel circuits are applied to components connected in parallel.

REVIEW QUESTIONS

Draw a sketch of each circuit before attempting solutions.

1. Four 40-ohm resistors are connected in parallel. What is the total resistance of the combination?

2. A 40-ohm resistor and a 20-ohm resistor are connected in parallel. The voltage across the 40-ohm resistor is 80 volts. What is the total current in the circuit?

3. A 10-ohm resistance is needed in a control circuit. Three 30-ohm resistors are available. How could these resistors be connected to obtain the required resistance?

4. A resistance is needed in a control circuit. The resistance must be more than 2 ohms but less than 4 ohms. Three 2-ohm resistors are available. How would you connect the 2-ohm resistors to obtain the required resistance?

5. Figure 5-20 is a lamp and switch circuit. Which switches must be closed to light the lamp, using the least number of switches?

6. In Figure 5-22, if the voltage across R_1 is 8 volts and the voltage across R_3 is 6 volts, the voltage across R_2 must be _____ volts.

7. Using the same voltages given in question 6, the generator voltage must be _____ volts.

8. In Figure 5-22, if the voltage from point A to point C is 18 volts and the voltage from point A to point B is 24 volts, then the voltage from point C to point B must be _____ volts.

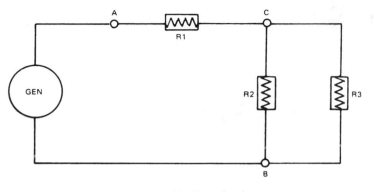

FIGURE 5-22 Test circuit.

9. In Figure 5-22, if the current through R_1 is 4 amperes and the current through R_2 is 3 amperes, then the current through R_3 must be _____ amperes.

10. In Figure 5-22, if the current through R_2 is 2 amperes and the current through R_3 is 3 amperes, and R_1 is a 10-ohm resistor, then the voltage across R_1 is _____ volts.

Unit **6**

Electric Meters

OBJECTIVES

Upon completion and review of this unit, you will have information on the following concepts:
- The construction of analog meter movements.
- The internal meter connections that provide for meter movement applications as an ammeter, voltmeter, or ohmmeter.
- In ammeter service, a shunt is used in parallel with the meter movement providing proportional amounts of current through the shunt and meter movements.
- In voltmeter service, a series resistor provides for current flow through the meter movement proportional to the voltage applied.
- When the meter movement is used to indicate resistance, an internal battery provides power while the resistance to be measured determines the current flow through the meter movement.
- The clamp-on ammeter provides a means of measuring current without the need of disconnecting wires.
- Digital meters provide an inexpensive rugged meter that is easy to read.

An electric meter is used to determine the characteristics of an electric circuit or component. The common types of meters used by the air-conditioning and refrigeration technician include the ammeter, voltmeter, and ohmmeter. Occasionally, the technician will have the opportunity to use a wattmeter. The meters used in the trade today include both analog and digital types.

ANALOG METER

The internal operation of analog meters is provided as a review of circuit operations. Series, parallel, and series parallel circuits provide for proper operation of analog meters.

An analog electric meter uses a meter movement that operates on the electric motor principle. To obtain meter pointer deflection, a current must pass through the meter-movement coil. This current sets up a magnetic field in the meter-movement coil that is repelled by the fixed magnetic field provided by the permanent magnets. The stronger the current flow through the meter-movement coil, the greater is the deflection of the coil and pointer, as shown in Figure 6-1.

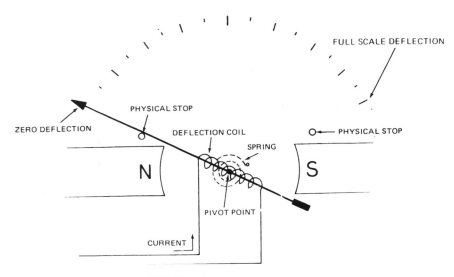

FIGURE 6-1 Meter movement.

In the meter movement shown in Figure 6-1, the permanent magnets are indicated by the capital N for north pole and capital S for south pole. The movement deflection coil is pivoted at the center and has a pointer attached to it. A coil spring attached to the meter movement holds the movement and pointer at zero (0) deflection when no current flows through the movement coil.

Physical stops in the form of pins are located just beyond the zero and full-scale deflection points to limit the movement of the pointer. When current flows through the meter-movement deflection coil, a magnetic field is set up around this coil. The magnetic field of the deflection coil will be repelled by the field of the permanent magnets. The meter movement will rotate around the pivot point, causing the pointer to move up scale.

As the meter movement rotates up scale, pressure will increase on the coil spring attached to the movement. The rotation will stop when the pressure of the repelling magnetic fields is balanced by the reverse pressure offered by the coil spring.

A different amount of current flow through the meter movement coil will cause a different amount of meter movement and pointer deflection.

AMMETER

An ammeter is used to measure current flow. Current flow is the movement of electrons through the circuit. To measure current flow, the circuit must be opened and the ammeter inserted so that the current to be measured flows through the ammeter. The circuit connections are shown in Figure 6-2.

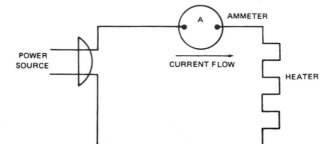

FIGURE 6-2   Ammeter connection.

Normally, the current to be measured is much higher than the amount needed to deflect the meter movement to full scale. To make meters useful in line-current measuring systems, a shunt is installed within the meter. The shunt is a low-resistance current path that allows a proportional amount of current to flow through the meter movement. The remaining current bypasses the meter movement by going through the shunt. This is shown in Figure 6-3.

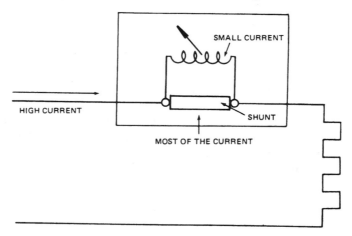

FIGURE 6-3   Shunt circuit.

Suppose that a 1-ampere full-scale meter were to be designed using a meter movement that required 0.001 amperes (1 milliampere, mA) full-scale deflection. The meter movement resistance is 100 ohms. According to Ohm's law, there would be 0.001 A × 100 = 0.1 V across the meter movement at full-scale deflection. The scale of the meter would be changed to indicate 1 ampere instead of 1 mA, but it would still take only 1 mA through the meter movement to cause full-scale deflection.

This arrangement is shown in Figure 6-4. One ampere is coming down the line. Only 1 mA goes through the meter movement, while 999 mA goes through the shunt. The shunt is in parallel with the meter movement. The resistance of the shunt is

$$\frac{0.1 \text{ V}}{999 \text{ mA}} = \frac{0.1 \text{ V}}{0.99 \text{ A}} = 0.1001 \ \Omega$$

The shunt in the ammeter is a low-resistance path for current flow. An ammeter is therefore a low-resistance device. An ammeter is *never* connected across the circuit. The low resistance of the ammeter will short the circuit, usually causing the line fuse to open. Ammeters are *never* connected in parallel with any component. Ammeters are *always* connected in *series*.

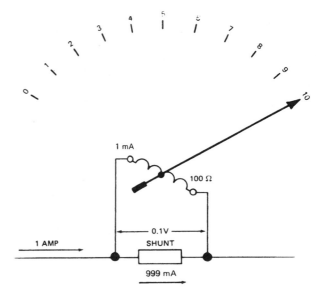

FIGURE 6-4 Shunt operation.

Many meters contain a switch system that provides more than one full-scale deflection sensitivity. A single meter could have a 1-, 10-, and 100-ampere full-scale reading. See Figure 6-5, Analog volt-ohm-milliammeter.

CLAMP-ON AMMETER

One problem in using a standard ammeter is that the circuit must be opened to insert the ammeter. In large, high-current situations, opening the circuit can be a time-consuming project. This is especially true if current is to be measured in a number of different current-carrying lines. The clamp-on ammeter was developed to simplify the current measurement procedure.

FIGURE 6-5   Volt-ohm-milliammeter. (Courtesy of BET, Inc.)

It was shown in Unit 2 that any wire carrying current has a magnetic field
about it. If the current is an alternating current (ac), the current will vary in the
form of a sine wave. The varying current will produce a varying magnetic field
about the wire. The strength of the magnetic field is directly proportional to the
strength or magnitude of the current flow.

The clamp-on ammeter uses a voltage step-up, current step-down transformer in the current-measuring process.

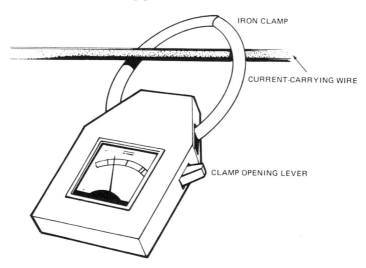

FIGURE 6-6   Clamp-on-ammeter.

An iron-clamping mechanism is placed around the wire carrying ac current,
as shown in Figure 6-6. Inside the clamp-on ammeter, the iron clamp completes
a magnetic circuit through a current transformer. The transformer produces an

output that is connected to a standard meter movement. The meter movement has a current scale that indicates the current flow in the ac current-carrying wire. After the current is measured, the clamp-opening lever is depressed, which opens the clamp, allowing the meter to be removed from the current-carrying wire.

A number of manufacturers provide clamp-on ammeters. The meters have different specifications, particularly with regard to current ranges. Always read equipment specifications before using the equipment.

VOLTMETER

A voltmeter can be developed with the same meter movement used in the ammeter. That meter movement required 1 milliampere for full-scale deflection and had 100 ohms of resistance.

In Figure 6-7, a 300-volt meter is shown. The scale has been drawn using 300 V as the full-scale deflection. Since voltage is measured across a circuit, the voltmeter will have to be placed across the circuit in order to measure voltage. There will be 300 volts at the terminals of the voltmeter. According to Ohm's law, the total resistance will equal the voltage divided by the current:

$$R = \frac{300}{0.001} \text{ or } 300,000 \text{ ohms}$$

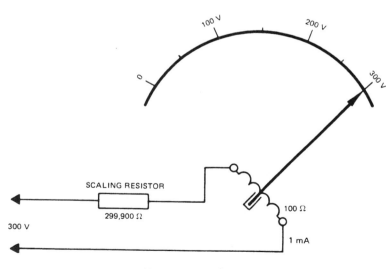

FIGURE 6-7   Voltmeter.

The scaling resistor would be 299,900 ohms, with the meter-movement resistance of 100 ohms added to it for the 300,000-ohm total. The procedure used in making a voltmeter is to determine the required resistance by dividing the desired full-scale voltage by the meter-movement full-scale current. The scaling resistor is then the total resistance minus the resistance of the meter movement.

OHMMETER

The ohmmeter will be developed using the same 1-milliampere full-scale meter movement. An ohmmeter requires the use of an internal power source. The standard source is a 1.5-volt cell. Observe the circuit shown in Figure 6-8. If the terminals of the meter were shorted together, a complete series circuit would exist. The power source is 1.5 volts. The series resistances are 100 ohms meter resistance, the 1300-ohm resistor, and the 300-ohm potentiometer. The potentiometer would be adjusted to provide a total resistance of 1500 ohms for the series circuit. Using Ohm's law, we then find

$$I = \frac{E}{R} = \frac{1.50}{1500}$$

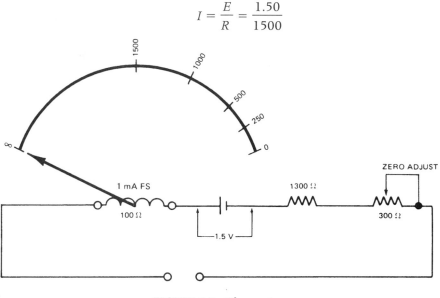

FIGURE 6-8   Ohmmeter.

The current flow through the circuit is 1 milliampere, the full-scale current of the meter movement. The meter movement rotates to full scale where the scale is marked zero (0) ohms. The meter leads are shorted together. The resistance is zero (0) ohms. The meter indicates 0 ohms.

If the meter leads were connected across a 1500-ohm resistor, the series circuit would contain the 1500-ohm resistor, the 100-ohm meter movement, the 1300-ohm resistor, and the potentiometer set at 100 ohms. The total resistance would be 3000 ohms. According to Ohm's law,

$$I = \frac{E}{R} = 0.0005 = \frac{1.5}{3000}$$

Thus 0.5 milliampere flows through the circuit. The meter movement would move to approximately half-scale, where the marking is 1500 ohms.

It is obvious that the value of the resistor connected between the terminals of the meter will determine the total current flow through the series circuit and therefore the amount of current through the meter movement. The current determines how far up scale the meter will rotate. A high resistance will keep the current low and provide little meter movement. A low resistance will provide a higher current and the meter may rotate close to zero (0) ohms. Many ohmmeters provide three or four multipliers for resistance measurement. Common multipliers include $R \times 1$, $R \times 10$, $R \times 100$, $R \times 1000$, and $R \times 10,000$.

DIGITAL METER

Many new meters are becoming available as new digital electronic circuitry is developed. Digital meters are easier to read than the standard meters since the meter reader has only to read the digits. With the older analog meters, the meter reader must interpolate, that is, decide what the meter is indicating. This process takes practice if correct answers are to be obtained. Figure 6-9 is a photograph of a digital multimeter.

FIGURE 6-9   Digital multimeter. (Courtesy of BET, Inc.)

SUMMARY

- Analog meters contain a meter movement (restrained electric motor).
- The same analog meter movement is often used to indicate current, voltage, and resistance.
- The connection of the meter movement to other electrical components, and the method of connection, determine whether current, voltage, or resistance will be measured.

• It requires experience to properly read an analog meter.

• A digital meter is not necessarily more accurate than an analog meter, although it is easier to read.

PRACTICAL EXPERIENCE

Required equipment Operable A/C system with access to leads to the compressor motor and condenser fan, clamp-on ammeter, and analog and digital voltmeters.

Procedure

1. In an operating system, measure the line voltage with an analog voltmeter. _____ volts.

2. Measure the same voltage with a digital voltmeter. _____ volts.

3. Which meter is more accurate? Discuss your answer to step 3 with colleagues.

4. Using the clamp-on ammeter, measure the current flow to the compressor motor. _____ amps.

5. Using the clamp on ammeter, measure the current flow to the fan motor. _____ amps.

6. For the purpose of steps 4 and 5, is the clamp-on ammeter easier to use than a standard ammeter?

Conclusions

1. The analog-meter movement is a spring-restricted small motor.

2. The analog meter responds to current flow through the meter movement coil.

3. In ammeter service, a shunt is used in parallel with the meter movement, providing proportional amounts of current through the shunt and meter movement.

4. In voltmeter service, a series resistor provides for current flow through the meter movement proportional to the voltage applied.

5. When the meter movement is used to indicate resistance, an internal battery provides power, while the resistance to be measured determines the current flow through the meter movement.

6. The clamp on the ammeter provides a means of measuring current without the need for disconnecting wires.

7. Digital meters serve as inexpensive rugged meters that are easy to read.

REVIEW QUESTIONS

The meter shown in Figure 6-10 is a volt-ohm-ammeter. The switch in the meter provides for the selection of *volts, current,* or *resistance.* The position of the switch selects the range of the meter.

The meter is shown with a pointer placed at different positions. Read the question, and indicate what value the meter is indicating.

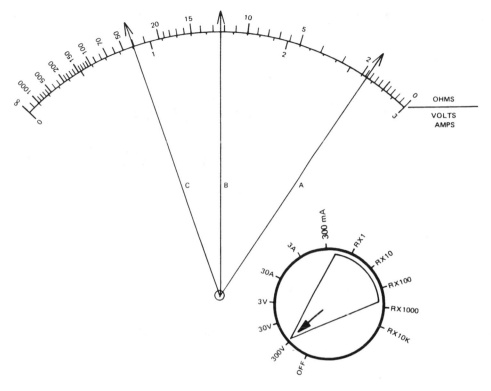

FIGURE 6-10

1. The selector switch is in the 30-V position. Pointer *B* position indicates _____ volts.
2. The selector switch is in the 3-A position. Pointer *A* indicates _____ amperes.
3. The selector switch is in the 3-V position. Pointer *C* indicates _____ volts.
4. The selector switch is in the 300-V position. Pointer *A* indicates _____ volts.
5. The selector switch is in the $R \times 10$ position. Pointer *A* indicates _____ ohms.
6. The selector switch is in the $R \times 1000$ position. Pointer *B* indicates _____ ohms.
7. The selector switch is in the 30-A position. Pointer *B* indicates _____ amperes.
8. The selector switch is in the $R \times 100$ position. Pointer *C* indicates _____ ohms.
9. The selector switch is in the $R \times 1$ position. Pointer *B* indicates _____ ohms.
10. The selector switch is in the 300-mA position. Pointer *A* indicates _____ mA.

Unit 7

Batteries and Electromotive Force

OBJECTIVES

Upon completion and review of this unit, you will
- Understand batteries as a source of direct current.
- Be able to demonstrate the relationship of balanced and unbalanced circuits.

BATTERIES

The purpose of this unit is not to explain how chemical energy is stored, converted, and then released as electrical energy when needed, but to cover the connection of cells, internal resistance, and the terminal voltage of batteries.

An Italian chemist-physicist, Count Alessandro Volta (1745–1827) invented the battery in 1780. His battery, the voltaic pile, was the first source of constant-current electricity. For his discovery, the *volt*, a unit of electromotive force, was named in his honor in 1881.

Whenever two different metals are placed in an acid or salts solution in an insulated case, a voltage is produced between the metals by chemical action. This combination, the two metals and the solution, is known as a *cell* (Figure 7-1). One plate is positive, and the other is negative. All cells and batteries provide a source of direct current.

The term *cell* refers to a single unit, the two metals in a solution. The term *battery* refers to cell combinations in series or in parallel that provide for higher current or voltage capabilities. A battery is used for storing and converting chemical energy into electrical energy. Commonly, the term battery is now used instead of cell; for example, a flashlight battery is actually a flashlight cell.

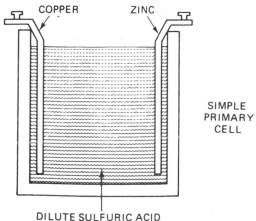

FIGURE 7-1 One cell.

There are two main types of cells: primary and secondary. *Primary cells* are temporary (nonrechargeable) and use up the materials of which they are composed while providing electrical energy, (Figure 7-2). *Secondary cells* change in chemical composition while providing electrical energy, but may be recharged. Recharging converts the cell back to its original condition by passing current through it in the opposite direction. A flashlight battery is a good example of a primary cell, and a car battery is a good example of a secondary cell.

FIGURE 7-2 Primary cells. (Courtesy BET, Inc.)

Chargers have become available that will recharge primary cells, including some flashlight batteries. Most batteries are not returned to their original condition, however. By design, some batteries are intended to be recharged, whereas others are intended to be discarded when discharged.

The symbol used for a cell is shown in Figure 7-3. The longer line is the positive terminal of the cell; the shorter line is the negative terminal.

The voltage output of a cell is determined by the materials used to make up the cell. The standard carbon-zinc (flashlight) cell comes in many sizes. The small penlight cell has an output of 1.5 volts. The common D size flashlight cell also has an output of 1.5 volts. The D size cell can, however, deliver higher current for a longer time than the penlight cell.

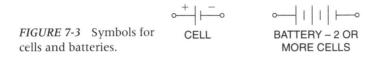

FIGURE 7-3   Symbols for   CELL   BATTERY – 2 OR
cells and batteries.   MORE CELLS

INTERNAL RESISTANCE

Everything through which current flows has resistance. Batteries have an internal resistance. If the open circuit voltage of a cell is measured and current is then supplied (by that cell) to an external circuit, its voltage will decrease. This decrease in voltage is equal to the voltage developed across the internal resistance of the cell. Figure 7-4 illustrates a battery with the internal resistance shown before the output terminals.

FIGURE 7-4   Battery or cell with
internal resistance.

A cell with a voltmeter in parallel, an ammeter in series, and a load resistor of 10 ohms in series controlled by a switch is shown in Figure 7-5. When the switch is open, the cell voltage is 1.5 volts. When the switch is closed, the ammeter indicates 0.14 ampere of current. The voltmeter indicates 1.4 volts. This means that 0.1 volt is being developed across the internal resistance of the battery (1.5 − 1.4 = 0.1). The internal resistance of the cell may be determined by using Ohm's law:

$$R = \frac{E}{I}$$

$$R = \frac{0.1}{0.14}$$

$$R = 0.714 \text{ ohm}$$

All batteries do not have the same internal resistance. As a cell becomes discharged, its internal resistance increases. As the internal resistance of the cell increases, its ability to supply current decreases.

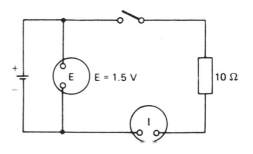

FIGURE 7-5 Determining internal resistance.

BATTERIES IN SERIES

Cells are connected in series whenever a voltage higher than that which can be supplied by a single cell is needed. Two flashlight cells are shown in series in Figure 7-6. The total voltage available is 1.5 plus 1.5 or 3 volts. This is standard for a two-cell flashlight. If three cells are connected in series, as in Figure 7-7, the available voltage is 4.5 volts (1.5 + 1.5 + 1.5 = 4.5).

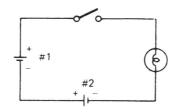

FIGURE 7-6 Two cells in series.

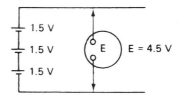

FIGURE 7-7 Three cells in series.

CELLS IN PARALLEL

Whenever higher current capability is needed, cells are connected in parallel. Certain precautions must be taken when connecting cells in parallel. Most importantly, the cells must have the same terminal voltage. For example, a 1.2-volt cell should *not* be connected in parallel with a 1.5-volt cell.

A load device drawing a constant current of 3 amperes requires a voltage of at least 1.3 volts (Figure 7-8). Cells of 1.5 volts are available. Each cell has an internal resistance of 0.1 ohm. With one cell connected to the load device, there is 0.3 volt developed across the internal resistance of the cell, leaving 1.2 volts to appear across the load. If two cells are connected in parallel, the situation is improved.

In Figure 7-9, the load device is still drawing 3 amperes. Each cell, however, is supplying only 1.5 amperes. The voltage developed across the internal resistance is

$$E = I \times R$$

$$E = 1.5 \times 0.1$$

$$E = 0.15 \text{ volt}$$

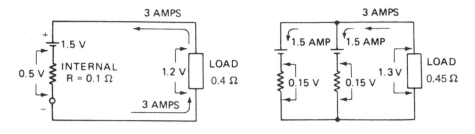

FIGURE 7-8   Internal voltage drop, single cell.

FIGURE 7-9   Lower internal voltage drop with parallel cells.

The voltage at the terminals of the cells and across the load is

$$E_{out} = 1.5 - 0.15$$

$$E_{out} = 1.35 \text{ volts}$$

This voltage meets the original requirement of at least 1.3 volts across the load.

Whether or not more cells should be connected in parallel depends on how long current is to be drawn. If three cells are connected in parallel, each has to supply only 1 ampere. The voltage developed across the internal resistance is only 0.1 volt. The terminal voltage of each cell and the voltage to the load is 1.4 volts (Figure 7-10). The three-cell combination will last longer than the two-cell combination.

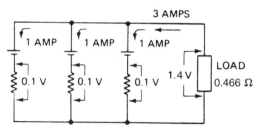

FIGURE 7-10   Further decrease in voltage drop with three-cell supply.

LINE CURRENT SPLITS IN 3 EQUAL PARTS

Internal resistance is present in every device that produces electricity, whether it is a battery, generator, bimetal strip, or solar cell. The effect of internal resistance is the same on these other devices as on batteries.

UNBALANCED AND BALANCED CIRCUITS

It has been shown that when two batteries are connected in series the total voltage available is the sum of the two battery voltages. The individual battery voltages are also available (Figure 7-11). This is similar to regular household electric power, where 240 volts ac may be available from the two hot lines, and 120 volts are available from either hot line to the neutral or ground wire (Figure 7-12).

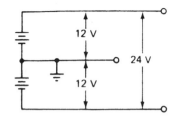

FIGURE 7-11 Two batteries
with ground.

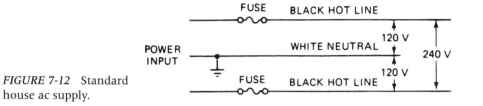

FIGURE 7-12 Standard
house ac supply.

Unbalanced Circuits

An example of an unbalanced circuit is shown in Figure 7-13. One resistor of 6 ohms is connected across a 12-volt section of the power source. A current of 2 amperes will flow in the circuit of battery *A* and in the 6-ohm resistor when there is no current flow in battery *B*.

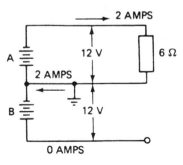

FIGURE 7-13 Unbalanced system.

If a 12-ohm resistor were added to the system, the circuit would still be un-balanced, as in Figure 7-14. Note that at the junction below the 6-ohm resistor, the current splits: 1 ampere returns to battery *A* while the other 1 ampere flows down through the 12-ohm resistor and through battery *B*. This 1 ampere joins with the 1 ampere returning on the neutral wire to provide the 2 amperes flow-ing through battery *A*.

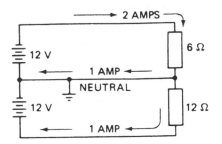

FIGURE 7-14   Unbalanced system, both lines loaded.

Balanced Circuits

If both resistors are of the same value, no current flows in the neutral wire (Figure 7-15). Although balanced circuits are possible under laboratory conditions, they are almost never obtained in actual practice in household situations. The unbalanced circuit is the most common.

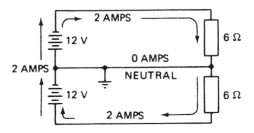

FIGURE 7-15   Balanced system.

The circuit in Figure 7-16 shows the lower voltage—as well as the higher voltage—components. In each junction, the current entering the junction is equal to the current leaving the junction. Ohm's law is true for each component of the system.

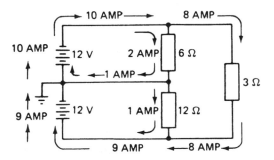

FIGURE 7-16   Unbalanced system using high and low voltage.

The relationship shown in Figure 7-16 parallels that of the 240/120 systems used in homes. The voltage levels in homes are different from those shown with batteries, but the circuit relationships are the same.

SUMMARY

- A cell is a combination of two dissimilar metals in an acid or alkaline solution.
- A battery is the combination of cells in series or in parallel.
- If cells are connected in series, the available voltage is the sum of the individual cell voltages.
- If cells are connected in parallel, the available current is the sum of the currents available from the individual cells.
- Batteries or cells should be connected in parallel only when they are of the same voltage.
- As a battery becomes discharged, its internal resistance increases.

PRACTICAL EXPERIENCE

Required equipment   Flashlight assembly, clip leads, and voltmeter.

Procedure

1. Remove the lamp portion of the flashlight.
2. Using clip leads, make connections completing the flashlight circuit.
3. With the light lit, measure the voltage across the lamp. _____ volts.
4. Disconnect the lamp assembly.
5. Measure the battery voltage. _____ volts.
6. Is the light *on* voltage lower than the light *off* voltage?

   NOTE: If an ammeter is available, the current flow (lamp on) may be measured. The internal resistance of the battery is equal to the no-load voltage minus the full-load voltage divided by the load current.

Conclusions

1. A balanced system is present when the current in both hot lines of a system is equal and there is no current in the neutral line.
2. In the home or in industry, the systems are more likely to be unbalanced than balanced.

REVIEW QUESTIONS

1. Whenever a higher current capability is needed, batteries are connected in _____.
2. When a higher voltage is needed, batteries are connected in _____.
3. Current flow through the internal resistance of a battery causes the output voltage to _____.

4. A battery has a no-load terminal voltage of 13 volts. When 1 ampere of current is drawn from the battery, the terminal voltage is 12 volts. What is the internal resistance of the battery?

5. As a battery becomes discharged, its internal resistance _____.

6. A 12-volt battery has an internal resistance of 0.5 ohm. How much current can be drawn from the battery before the terminal voltage decreases to 10 volts?

7. How many 1.5-volt batteries must be connected in series to obtain 13.5 volts?

8. In a balanced circuit, _____ current flows in the neutral wire.

9. Is the balanced or unbalanced circuit more common in actual practice?

10. Refer in Figure 7-16. The voltage across the 3-ohm resistor is _____ volts.

Alternating Current

OBJECTIVES

Upon completion and review of this unit, you will have learned of

- The development of current and voltage by mechanical means (generator-alternator).
- The law regarding the movement of electrons when passing through a magnetic field.
- The generation of voltage and current in rotating coils and rotating magnetic fields.
- The efficiency of polyphase generators.
- The effects of wye and delta connections.

In the previous discussion of electrical current, the implication was that electrons always move in one direction. Actually, the most common form of electrical energy generated today is *alternating current,* ac, which changes direction 60 times each second.

The standard frequency in the United States is 60 cycles per second or 60 hertz (Hz). Another common frequency that is used in some countries is 50 hertz.

Alternating current is the most common form of electrical energy because it is the natural form of a generated current. Alternating current may be delivered over great distances with much less loss in power than direct current (dc). For these reasons, alternating current is generally less expensive than direct current for most applications.

GENERATION OF AC VOLTAGE

In Unit 2, we discussed the fact that an electric current flowing in a wire produces a magnetic field about the wire. Another relationship between magnetic fields and electricity is the following:

Whenever a wire passes through a magnetic field, a voltage is induced in the wire.

The amount of voltage induced in the wire depends on two factors: the strength of the magnetic field and the speed at which the wire cuts through the field.

In generating electric voltages,

Stronger magnetic field = higher voltage

Higher speed = higher voltage

Generated Electromotive Force (Voltage)

A *generator* is a device that converts mechanical energy into electrical energy by mechanically rotating coils of wire in a magnetic field. An *alternator* is a device that converts mechanical energy into electrical energy by rotating magnetic fields past coils of wire.

In Figure 8-1, a piece of wire is shown moving down through a uniform magnetic field. In position 1, the wire is not cutting through the field, and no voltage will be generated. As the wire enters the magnetic field, position 2, a level of voltage is established. As the wire moves through the uniform field, position 3, at a constant speed, the level of voltage remains constant. At position 4, the wire leaves the magnetic field, and the voltage drops to zero. In position 5, the wire does not cut through the magnetic field, and the voltage remains at zero.

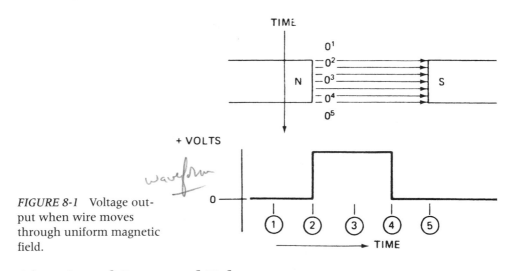

FIGURE 8-1 Voltage output when wire moves through uniform magnetic field.

Direction of Generated Voltage

When a piece of wire passes through a magnetic field, some of the free electrons in the wire are forced to move. The direction in which the electrons move, and therefore the polarity of the voltage, depends on two things:

1. The direction in which the wire moves through the magnetic field.
2. The direction of the magnetic field.

Left-Hand Rule

There is a rule associated with generator action, the left-hand rule. The rule states: receive the lines of force in the palm of the left hand with the thumb pointed in the direction of wire motion. The extended fingers will point in the direction of electron movement. (Remember, lines of force travel from the N pole to the S pole.)

Figure 8-2 shows an example of the rule. The direction of electron movement is indicated in Figure 8-2 with a (+) at the tail of an arrow. The electrons move from the (+) end of the wire to the other end, creating a voltage. The (+) end is positive, since electrons move away from this end of the wire. The other end of the wire is negative because the electrons crowd in.

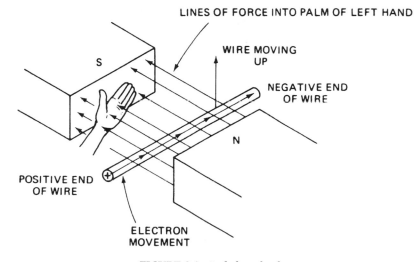

FIGURE 8-2 Left-hand rule.

Rotating Coils

For practical mechanical reasons, generators are made up of coils rotating in a magnetic field. When a coil rotates in a magnetic field, an alternating voltage is produced. In Figure 8-3, an example of coil section, one wire is shown. As the wire rotates through a complete revolution, a complete cycle of alternating voltage is produced. The complete revolution is shown, divided into 12 equal parts.

In position 0, the wire is moving parallel to the magnetic lines. The wire does not cut through them. No voltage is produced. As the wire moves from position 0 to position 1, it cuts through a few magnetic lines. A (+) indicates the direction of electron movement and produced voltage. The graph, Figure 8-4, indicates the level of voltage. As the wire moves from 1 to 2, it cuts through an increasing number of lines. A corresponding increase in voltage is indicated on the graph. At position 3, the wire is cutting directly down through the magnetic lines. The

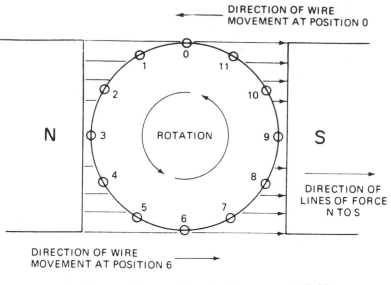

FIGURE 8-3   Rotating wire in a magnetic field.

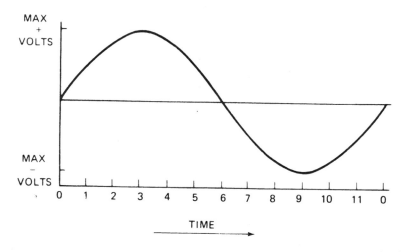

FIGURE 8-4   Sinewave generation from rotating a wire in a magnetic field.

maximum voltage is generated, as indicated in the graph. As the wire moves to positions 4 and 5, it cuts through fewer lines, until at position 6, it is again moving parallel to the magnetic lines. In the graph, this is shown as decreasing voltage to a zero value at position 6. As the wire moves from position 6 to position 7, the direction of movement through the magnetic lines changes.

The level of voltage starting at 7 is increasingly negative, reaching a maximum at position 9. The voltage decreases at 10 and 11, falling to zero volts at position 0. This completes one revolution and one complete cycle of voltage. As the wire continues to rotate, another cycle starts.

According to the left-hand rule for positions 1 through 5, electron motion would be into the page as the wire moves down. In positions 7 through 11, the wire is moving up. The left-hand rule provides for electron motion away from the page, as indicated by a dot in the center of the wire symbolizing the point of an arrowhead.

The voltage is called *alternating voltage* because the output of the generator will first be positive and then negative at each output terminal.

Most power companies in the United States generate electricity with a frequency of 60 hertz (Hz) or 60 cycles per second (cps). There are 60 complete cycles of positive and negative alternations during each second. Equipment designed to operate with voltage at this frequency is marked on the nameplate at 60 cps or 60 Hz. Many foreign countries produce voltages and operate equipment at 50 cycles per second (50 cps or 50 Hz).

Polyphase Generators and Alternators

Generators that produce a single output, alternating voltage are called single-phase generators (Figure 8-5). The voltage completes one cycle (360°) in $\frac{1}{60}$ second, for standard U.S. supply.

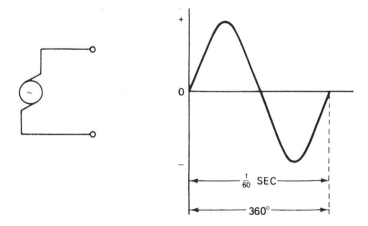

FIGURE 8-5 Standard U.S. 60-cycle voltage.

A second generator could be connected to the same drive source and produce an output voltage (Figure 8-6). If the armatures of the two generators are exactly aligned, as in Figure 8-7, the output voltages will be in phase with each other.

Another alignment of the armatures could be made that would not provide in-phase voltages at the outputs. In Figure 8-8, the armature of generator B has been rotated clockwise 90°. The output of generator B is 90° out of phase with the output of generator A. The two outputs of the generators could be connected together (Figure 8-9). The voltages available would then be as given in Figure 8-9.

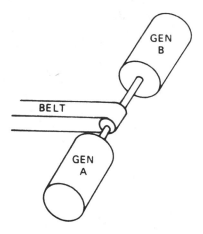

FIGURE 8-6 Two generators operating at the same speed.

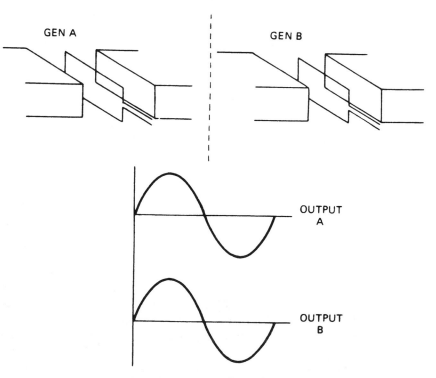

FIGURE 8-7 Two in-phase sine waves.

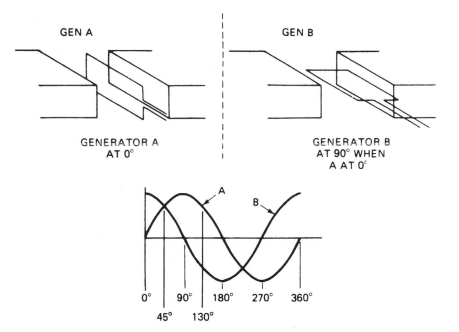

FIGURE 8-8   Two sine waves of voltage 90° out of phase (two phase).

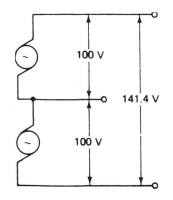

FIGURE 8-9   In the two-phase system the line voltage is not the direct sum of the individual voltages.

The output between the two generators is not the sum of the two voltages because the voltages are not in phase and cannot be added directly. Since the voltages are 90° out of phase with each other, the output between A and B will be 100 times the square root of 2 ($\sqrt{2} = 1.414$).

$$100 \times 1.414 = 141.4 \text{ volts}$$

SUM OF THE INSTANTANEOUS VALUES

Consider the addition of instantaneous values of voltages at intervals throughout the cycle. In Figure 8-8, phase A is 0 volts at 0 degrees, while phase B is at maximum, 100 volts at 0 degrees. The sum of the two voltages at this instant is 100 volts and is shown in Figure 8-10 as the 0° voltage.

At 45°, both phases A and B are 70.7 volts. The sum of the two voltages is 141.4 volts. This is shown as the peak positive voltage at 45° in Figure 8-10. If the procedure through 360° is continued, the waveform shown as Figure 8-10 will be completed. Note that the phase angle of the waveform in Figure 8-10 is halfway between phases A and B of Figure 8-8.

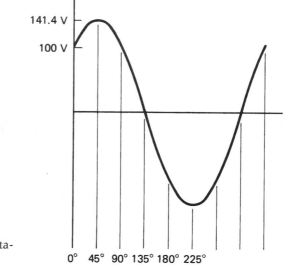

FIGURE 8-10   Sum of instantaneous values

The sum of the two voltages in Figure 8-8 is a voltage, 141.4 volts peak. This voltage lags phase B by 45° and leads phase A by 45°.

If two-phase power is required, a single generator could be constructed with armature coils placed 90° out of phase with each other (Figure 8-11). The output of the generator with windings 90° out of phase with each other would be the same as that shown in Figure 8-8: the voltages would be 90° out of phase with each other. Two-phase generators or alternators are not very common.

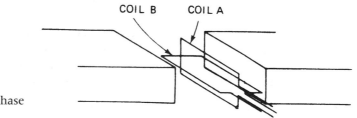

FIGURE 8-11   Two-phase generator.

In actual practice, it is more practical to have three rather than two windings in an alternator. Most ac power is produced in alternators since they are easier to construct and do not require brushes for the high-power output. The space in the alternator is put to almost complete use. The load that is presented to the prime mover, whether it is a steam turbine or water pressure from a dam system, is more constant.

In three-phase alternators, the windings are placed 120° out of phase with each other. The alternator output voltages produced are shown in Figure 8-12.

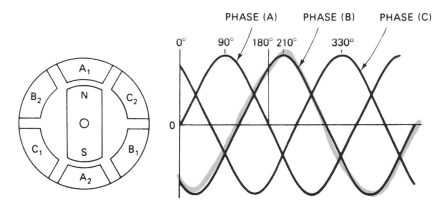

FIGURE 8-12 Three-phase alternator

Wye and Delta Connections

There are two methods of connecting the windings of alternators; each method produces different output voltages. They are called the *wye* and *delta connections*. The wye connection provides for higher voltage output, whereas the delta connection provides for higher current capabilities.

The output windings of the alternator in Figure 8-12 could be connected as shown in Figure 8-13. The common connection could be connected to ground and could provide a neutral wire. If each winding is producing 120 volts, 208 volts is available between each pair of phase wires. In the wye connection, the output voltage is equal to the winding voltage times the square root of 3 ($\sqrt{3} = 1.732$).

$$120 \times 1.732 = 208$$

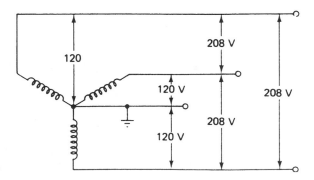

FIGURE 8-13 Wye connection.

The voltage between any output phase wire and neutral would be the winding voltage of 120 volts.

In the delta connection of windings, the output voltage is equal to the winding voltage (Figure 8-14). There are 120 volts between each of the three-phase output lines. Current flow in a delta system splits. If load resistors of 6.93 ohms (Figure 8-14) are connected between the output lines, the current flow is

$$I = \frac{E}{R}$$

$$I = \frac{120}{6.93}$$

$$I = 17.32 \text{ amps}$$

This is the current in each resistor.

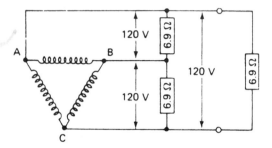

FIGURE 8-14   Delta connection.

When the current in a line reaches the alternator terminals, it splits to two windings. Each winding has only 10 amperes flowing in it. In the delta connection, the winding current equals the line current divided by the square root of 3.

$$I_{\text{winding}} = \frac{I\,\text{line}}{\sqrt{3}}$$

$$I_w = \frac{17.32}{1.732}$$

$$I_x = 10 \text{ amps}$$

The three-phase system of producing electric power is the most common in the world today.

EFFECTIVE VALUE OF VOLTAGE AND CURRENT

When a dc voltage is applied to a resistor, a level of current is established, and it flows in the circuit. In a dc circuit, the effects are constant (Figure 8-15).

In an ac circuit, the voltage is varying continuously. When an ac voltage is applied to a resistor, varying amounts of heat are produced as the voltage varies (Figure 8-16). Although the applied voltage varies in both the positive and nega-

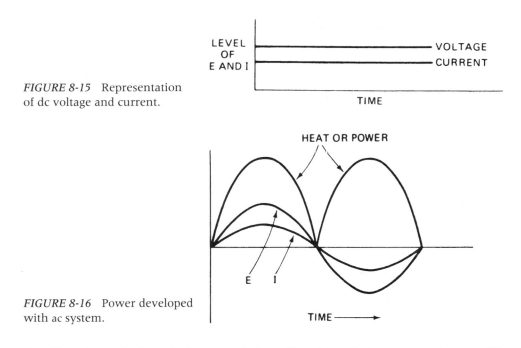

FIGURE 8-15 Representation
of dc voltage and current.

FIGURE 8-16 Power developed
with ac system.

tive directions, the heat is shown as being all positive. It does not make any difference in the resistor which way current flows through it; the result is heat.

In Figure 8-16, it is obvious that the same amount of heat is not produced at all times: there are peaks and valleys. At the 0°, 180°, and 360° marks, no heat is produced. The overall result is that if an alternating voltage of 100 volts peak is applied to a resistor, it will have the same heating effect as 70.7 volts dc. By formula,

$$\text{Effective value} = 0.707 \text{ peak value}$$

The effective value of voltage is sometimes called the *rms* value. The abbreviation *rms* stands for root mean square. In later units, the relation between the heat developed in a resistor and the square of the current value will be developed.

Voltmeters used to measure electric voltages are designed to indicate the effective, or rms, value of voltage. The air-conditioning technician will have little cause for concern with peak values. This presentation on peak and effective (rms) values was given for information purposes only.

SUMMARY

- A voltage is induced in a wire whenever the wire is passed through a magnetic field.
- The motion of an electron in a magnetic field is covered by the left-hand rule.
- In a generator, voltage is induced in a rotating armature as the armature rotates in a magnetic field.

- In an alternator, voltage is induced in a fixed armature through the use of a rotating magnetic field.
- In both a generator and an alternator the induced voltage is in the form of a sine wave.
- In many generators and alternators three sets of coils are used in the armature. The induced voltage is called *three-phase* voltage.
- Three-phase connections are wye, delta, or open delta.

PRACTICAL EXPERIENCE

Required equipment   Voltmeter (ac), and access to three-phase voltage panel.

Procedure

1. Using the voltmeter, measure the voltage between the three phases of voltage.

    Phase 1 _____ volts.

    Phase 2 _____ volts.

    Phase 3 _____ volts.

2. Measure the voltage from each phase to ground.

    Phase 1 _____ volts.

    Phase 2 _____ volts.

    Phase 3 _____ volts.

3. Is the transformer supplying the system connected in wye or delta?

4. If this is a delta system, identify the high phase.

Conclusions

1. Whenever electrons are in motion in a magnetic field, the magnetic field will distort the direction of the original motion of the electrons.

2. Because of this relationship of magnetic fields and electrons, the generation of voltage and current is possible.

3. Practical forms of electromechanical generators of electricity provide alternating current.

4. Three-phase generators may be connected in either the delta or wye configuration.

REVIEW QUESTIONS

1. What is the most common form of electrical energy in the United States?

2. What is the frequency of the ac produced in the United States?

3. What two factors determine the voltage induced in a wire?

4. What is the left-hand rule for generator action?

5. The abbreviation dc stands for _____.

6. The abbreviation ac stands for _____.

7. In the two-phase system, the voltages are out of phase by _____ degrees.

8. In the three-phase system, the voltages are out of phase by _____ degrees.

9. In the wye system with a winding voltage of 100 volts, the line voltage is _____.

10. In the delta system with a line current of 34.4 amperes, the winding current is _____ amperes.

Unit 9

Electrical Safety

OBJECTIVES

Upon completion and review of this unit, you will understand
• Effects of current flow through the body.
• Factors affecting the amount of current flow.
• Dangerous conditions and equipment.
• First aid for shock victims.

EFFECTS OF CURRENT FLOW THROUGH THE BODY

Many tests have been made attempting to determine the effects of current levels on the human body. The general effects are shown in Table 9-1. Keep in mind that individuals will have different responses to current flow through their bodies. The chart, then, shows the approximate amount of current causing a given body response.

Besides the responses given in Table 9-1, the secondary reactions to electrical shock are often the really serious part of the encounter. Any electrical shock that causes an involuntary response can cause a serious accident.

TABLE 9-1

Amount of current (mA)*	Response
0–0.5	No response
0.5–2	Slight tingling to mild shock; quick withdrawal from body contact
2–10	Mild to heavy shock; muscular tightening
10–50	Painful shock; cannot let go
50–100	Severe shock; breathing difficulties
100–200	Heart convulsion; death
Over 200	Severe burns; breathing stops

*1 mA = 0.001 ampere.

For example, a technician might be on a ladder or scaffolding while using a drill. If she or he receives a shock, the technician might involuntarily drop the drill. This could cause a serious accident to anyone who happened to be below the drill.

Another example is a technician working on a roof of a building. If she or he were to receive a strong electrical shock, a combination of voluntary and/or involuntary muscle contractions could cause the technician to jump off (be knocked off) the roof.

She or he might live through the shock, but the fall from the roof might be another matter.

It is important that the technician always be aware of the electrical dangers faced in the service area.

FACTORS AFFECTING BODY CURRENT FLOW

It was stated earlier in this textbook that Ohm's law is true for the whole circuit or any part of the circuit. When the human body becomes the circuit (the path for current flow), Ohm's law is true. The factors needed in determining the amount of current are the applied voltage and the body resistance. The applied voltage is the voltage with which the body comes in contact. The resistance of the body is the variable factor.

If the resistance of the human body were high enough, for example, 1 million (1,000,000) ohms, electrical shock would not be a problem.

EXAMPLE _____

Air-conditioning and refrigeration service technicians normally work with ac voltages under 600 volts. If the body resistance were 1 million ohms, the current flow by Ohm's law would be

$$I = \frac{E}{R}$$

$$I = \frac{600}{1,000,000} = \begin{array}{l} 600 \text{ microamperes} \\ \text{or } 0.6 \text{ milliampere} \end{array}$$

The 0.6 milliampere of current is barely enough to cause a human to respond or even recognize that a shock was received.

The problem is that the resistance of the human body is not a fixed value. In addition, the resistance of the human body is mainly the skin resistance. Internally the body is mostly water, and salty at that. Salt water is a good conductor of electricity. It has low resistance.

Since it is the skin resistance that controls the overall body resistance, let's take a closer look at skin resistance. Dry skin has a resistance of about 200,000 to 700,000 ohms. A resistance at this level would be sufficiently high, keeping current flow below a value that would cause a problem.

Wet skin is a different situation. The resistance of skin wet with perspiration or other moisture can go as low as 300 ohms. Any contact with line voltage when your skin is wet can cause serious injury or death. Remember, Ohm's law is always true.

$$I = \frac{E}{R}$$

$$I = \frac{120 \text{ volts}}{300 \text{ ohms}}$$

$$I = 400 \text{ milliamperes}$$

According to Table 9-1, a current flow of 400 milliamperes through the body can cause death.

Consider the conditions in which a service technician might be working. One area in which a technician is often required to work is a kitchen. Here the high temperature can easily cause a body to perspire. In a short time, the body, as well as shirt, pants, socks, and shoes, become damp with perspiration. Body resistance, as well as contact resistance through clothes, would go down rapidly. Any contact with normal working voltages could result in dangerous shock and burns.

Keep in mind that dry clothes provide good insulation. Clothes wet with perspiration or other moisture are usually low-resistance paths for electricity.

When you are wet, it is dangerous to come in contract with line voltages.

DANGEROUS WORKING CONDITIONS AND EQUIPMENT

As stated earlier, a dry body in dry clothes and dry shoes has sufficient resistance so that contact with a live wire should not cause a dangerous shock. The major difficulty is that it is nearly impossible to maintain these conditions when working. The dangerous working condition is to work around *live* line voltages whenever your body is damp with perspiration or other moisture. It is also dangerous to work around wet equipment, where a better contact to the body may be made through wet metal surfaces. Whenever water is present, whether it be a pool of water on a roof or floor or dampness due to perspiration or moisture on metal cabinets, consider the situation dangerous from an electrical standpoint.

HAND TOOLS

The service technician often encounters problems where power hand tools must be used. Whenever a power hand tool is picked up, line voltage (120 volts ac) is close to the gripped hand tool. This should always be considered a potentially dangerous situation.

The first rule in using power hand tools is a matter of common sense but is often ignored until someone receives a dangerous shock.

1. If anyone receives a slight tingling shock while using a power hand tool, the tool should be discarded *immediately.* Do not suggest that the job be finished and then the tool replaced. Stop using the tool at the first sign of a shock.

The second rule regarding power hand tools is an extension of the first rule.

2. When discarding electrically dangerous power hand tools, destroy the hand tool completely, thus ensuring that an unsuspecting individual does not find the tool and receive a dangerous or fatal shock while attempting to use it.

The electrical safety problem with power hand tools is always present when the older model, two-wire, metal-case power tool is used. The same problem exists when the three-wire power hand tool is used where a good ground does not exist. When a shock occurs, the electrical circuit causing the possibly dangerous shock must first be investigated. The lack of good grounding with the three-wire, grounded case tool will then be looked into.

In Figure 9-1 a hand-drill motor with a motor winding short to the metal case is indicated. A stick diagram of the individual holding the tool is also shown. This is a diagram of a dangerous situation. A number of different results could be obtained from this situation. Let's consider the two extremes.

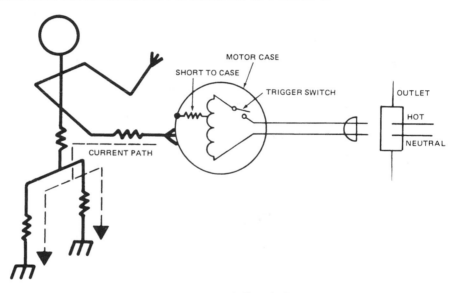

FIGURE 9-1   Two-wire drill with short to case.

1. The work location is dry; the technician is dry; this is the first task of the day. The technician's hand's resistance to the drill case is high, and grounding through the technician's left hand and feet is through high resistance.

   This situation could result in the technician not feeling any shock at all or perhaps a slight tingling shock. Technician would be lucky to feel the slight tingling shock and therefore know that the drill was dangerous and should be discarded. Otherwise, the drill might be continued in use, resulting in situation 2 at some later time.

2. The work location is damp; the technician is wet with perspiration. The work site is cramped between two damp metal cabinets, one of which needs a hole drilled through its case. Since the metal case of the cabinet to be drilled is

thick, pressure will be needed on the drill. If the technician braces his or her back against the case of the second cabinet and presses the hand trigger switch on the shorted drill, the chances are the technician will *die* right there.

It is obvious that if the drill were repaired after the first indication of a tingling shock, the technician would not have been killed in the second situation.

GROUNDING OF TOOL HOUSING

If the metal case of a three-wire hand tool is properly grounded, a short within the tool will most likely result in a blown fuse. Even if the fuse does not blow, the possibility of a dangerous electrical shock is lessened to a great extent since the case of the tool will be at close-to-ground potential, as shown in Figure 9-2.

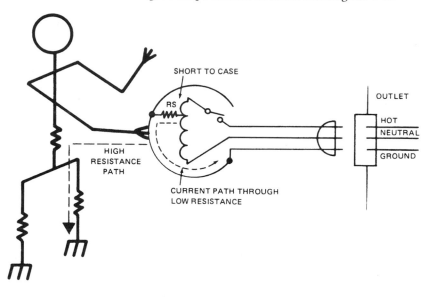

FIGURE 9-2   Three-wire drill with short to case.

One of the main points for electrical safety when using hand tools with three-wire power cords is to ensure that the ground connection is complete to ground. One problem encountered with three-wire power cords is to find the grounding prong broken off. This usually occurs when someone has previously used the hand tool where only the old type of two-wire receptacle was available. The hand tool should not be used until a new three-wire plug or complete cord has been installed.

Cheater plugs adapting the old two-hole receptacles to the three-wire plug are available. The ground pigtail must be connected to a good ground if a safe situation is to exist. Do not be satisfied with a connection to a screw holding the outlet cover plate. Make sure the outlet box is grounded. The use of double-insulated tools is making inroads in the air-conditioning field. Since the tools use the two-prong plug, the need for the three-prong grounding receptacle is eliminated. The problems of the ungrounded three-prong receptacle are also eliminated.

Since safety engineers have not yet come to a definite conclusion as to double-insulated tools, it would be well to keep abreast of this changing situation.

Panel taps are often used on job sites as a source of 120-volt power whenever standard outlets are not present. An example might be a rooftop installation. An example of a panel tap is given in Figure 9-3.

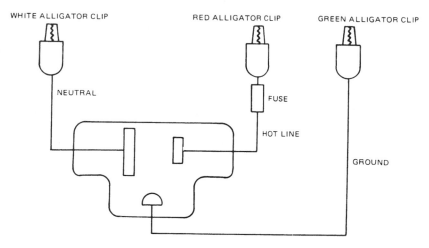

FIGURE 9-3   Three-wire cheater plug.

The panel tap is made from a standard three-wire extension cord with a molded receptacle at one end. The plug is removed from the opposite end and the wires separated for approximately 2 feet. The clips are installed at the wire ends as shown in Figure 9-3. An insulated in-line fuse cartridge is installed in the red line. Maximum fuse size is 15 amperes. The panel tap is practical if properly constructed and installed.

First, inspect the panel tap to ensure that the clips are properly fastened and color coded. Check the panel power with a voltmeter for a source of 110 to 120 volts to neutral/ground at the panel box. Do not assume the voltage is 120 volts to neutral/ground. You may have a 440- or 660-volt power source.

When the nominal 120-volt source has been located, panel tap connections may be made. The panel tap extension cord should be tied off, providing strain relief at the power box. Movement of the cord at the receptacle end should not cause movement at the clip end. Next, make the ground connection. If a ground bus bar is available, the *green* ground clip should be connected to it. If there is no ground bus bar, the *green* clip should be connected to the metal case of the box or metal conduit. Make sure that paint does not isolate this connection.

The second connection to be made is the neutral *white* clip. This clip should also be connected to the neutral/ground bus, if available, or to the metal case of the power box. The *red* clip may then be carefully connected to the hot-line terminal. If practical, these connections should be made in a power-off situation.

The technician creates a somewhat unsafe condition when using a panel tap. It should be recognized as such and used only when absolutely necessary.

FIRST AID FOR ELECTRIC SHOCK

In cases of electric shock, the first and most important step is to disconnect the victim from the electric power source. If possible, turn the power off. If this is not possible, use a dry board or other nonconducting material to separate the victim from the electric connection.

Start artificial respiration as soon as it is safe to touch the victim. Speed is essential when dealing with victims of electric shock. Do not stop the administration of the artificial respiration even if the victim appears dead. Long time periods of up to 8 hours have elapsed before some victims have responded. Only a physician should determine if the victim has died.

Most importantly, technicians should take the official first-aid course offered in their areas. It is extremely important to know the best methods of administrating first aid to shock victims.

Keep yourself aware of accidents as they happen by reading your trade periodicals and newspapers. A recent accident reported in a trade paper described a situation where a technician was working in the attic of a manufacturing plant. Being safety conscious, he had disconnected power from the air-conditioning system he was working on. The attic was not well lighted, it was hot, he was perspiring, and his clothes were damp. Along with the electrical conduit feeding the air-conditioning systems were commercial gas lines providing service to the building.

The technician had just finished placing a blower motor in the air distribution system. He raised himself up and possibly stretched to relieve the tension in his back. At his head level behind him was an open junction box. The cover had been left off by a previous worker. A wire nut had loosened and fallen off a pigtail connection, or perhaps it had never been installed.

As the air-conditioning technician stretched back, his head came in contact with the hot pigtail. The pigtail was the junction in a 278-volt lighting system. The air-conditioning technician had his hand and back against the gas line, which is an excellent ground. The attempts to revive him were not successful.

Be aware of what you are doing, where you are, and what is around you. Be safety conscious at all times. Do not be satisfied with a half-hearted approach to an understanding of safe working conditions.

SUMMARY

- With electrical shock, it is the amount of current flow that determines the effect.
- Skin resistance is the main control of body resistance. Wet skin means low resistance. Low resistance means high current. High body current is very dangerous.
- If you feel a shock when using a hand tool, do not use the tool again or allow for any other person to use that tool.
- Always be aware of safety considerations. Your life depends on it!

REVIEW QUESTIONS

1. The only electrical shocks that are dangerous are the ones that cause harsh burns.　　　　T_____　F_____

2. A victim of electric shock is always knocked away from the electric connection by involuntary muscle contractions.　　　　T_____　F_____

3. Most shocks are the result of carelessness.　　T　　　　F_____

4. No voltage should be considered safe, since it is the current that causes damage, and body resistance is changeable.　　　　T_____　F_____

5. Hand power tools are not dangerous if grasped with only one hand.　　　　T_____　F_____

6. A slight tingling shock from a hand tool is not important.　　　　T_____　F_____

7. A technician can feel safe as long as a three-wire plug is on his or her hand drill.　　　　T_____　F_____

8. Two wire connections are needed on a panel tap.　T_____　F_____

9. The voltage between the black wire in a panel box and ground is 120 volts.　　　　T_____　F_____

10. In case of electric shock, the first thing to do is to remove the victim from the electrical connection.　T_____　F_____

Unit 10

Capacitance and Inductance

OBJECTIVES

Upon completion and review of this unit, you will know that
- Capacitance relates to the storage of electric energy in electrostatic fields.
- Inductance relates to the storage of electricity in magnetic fields.
- Capacitance opposes a change in voltage.
- Inductance opposes a change in current.
- Current and voltage in an ac circuit are not always in phase with each other.

The first hint of capacitive action was given in Unit 1. A generator was shown pulling electrons from a wire connected to the positive terminal of the generator and pushing them out on the wire connected to the negative terminal. A similar situation is shown in Figure 10-1; the generator is running and wires are connected to the positive and negative terminals. For a voltage to appear between the positive and negative wires, electrons have to be pulled from the positive side and forced out on the negative side.

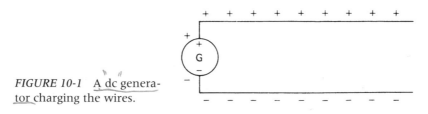

FIGURE 10-1 A dc generator charging the wires.

If a circuit constructed with a sensitive meter were connected as shown in Figure 10-2, the meter would indicate electron movement or current flow during the charging of the wires. The meter needle would move up as the switch is closed

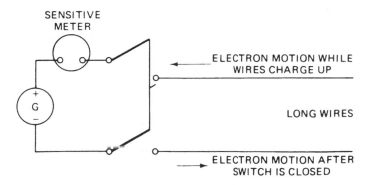

FIGURE 10-2   A meter showing charging current when switch is closed.

and immediately drop back to zero as the wires became charged. The number of free electrons that move would control the meter needle movement.

The number of electrons that have to move to "charge" the wires could be increased by replacing the external wires with large, flat plates. Increasing the surface area enlarges the charge capability.

Metal plates with a large surface area separated by a high-resistance insulating material make up a capacitor. In Figure 10-3, air is the insulating material.

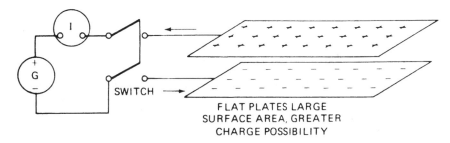

FIGURE 10-3   Increased charging current when large plates are used.

DC CURRENT FLOW IN CAPACITIVE CIRCUITS

In the explanation for Figure 10-2, it was indicated that after the switch is closed there is a momentary movement of electrons. With the dc generator connected to the wires, some of the free electrons move from the top wire through the generator and out on the bottom wire. After the wires become charged, voltage appears between all points on the top and bottom wires. Electron movement then stops.

A similar situation exists if a capacitor is connected across a dc generator or any other dc power source. As the capacitor is connected across the generator, the current jumps up. As the capacitor charges, the current flow drops to zero and the source voltage appears across the capacitor terminals. A graph of voltage and current as related to time is shown in Figure 10-4.

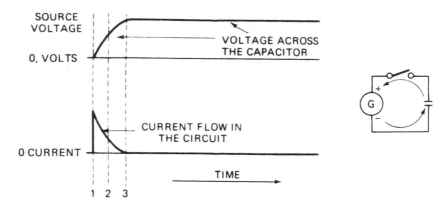

FIGURE 10-4   Current to and voltage across a capacitor with time.

Current

1. As the switch is closed at time zero, the current rises immediately to maximum.
2. The current decreases as the capacitor charges.
3. The capacitor is fully charged; the current drops off to 0 amperes.

Voltage

1. As the switch is closed, the voltage across the capacitor rises rapidly from 0 volts.
2. The voltage rise starts to taper off as the voltage across the capacitor approaches the source (generator) voltage.
3. The voltage across the capacitor reaches the source voltage; the capacitor is fully charged. The voltage across the capacitor remains constant at the source voltage level.

AC CURRENT FLOW IN CAPACITIVE CIRCUITS

When an ac generator is connected across a capacitor, the relation of current and voltage is similar to the situation with a dc source. The current and voltage relationship with an ac source is shown in Figure 10-5.

1. At time position 1, the voltage is zero and current flow is maximum (Figure 10-6a).
2. At time position 2, the voltage across the capacitor is rising and the current flow is decreasing (Figure 10-6a).
3. At time position 3, the voltage is maximum while the current is at zero.

    In steps 1 through 3, current was flowing into the capacitor while the capacitor was charging up. This is shown in Figure 10-6a. Electrons move into the bottom plate of the capacitor and out of the top plate.

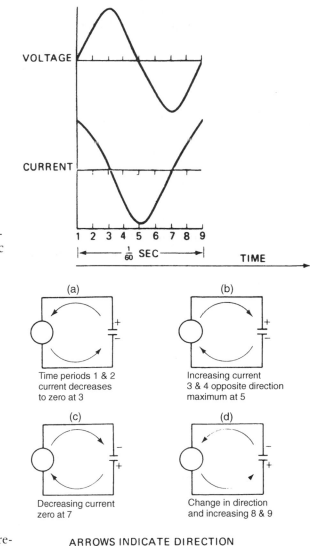

FIGURE 10-5 Current and volt-age are 90° out of phase with ac voltage; current leads voltage.

FIGURE 10-6 Current flow, as re-lated to Figure 10-5.

ARROWS INDICATE DIRECTION OF ELECTRON MOVEMENT

The voltage across the capacitor is increasing. As step 3 is passed, the current flow changes direction. Electrons start to move back out of the bottom plate and into the top plate of the capacitor. This is shown in Figure 10-6b. The voltage across the capacitor is decreasing.

4. At time position 4, the voltage across the capacitor decreases as the current flow increases.

5. At time position 5, the current flow is maximum. The voltage across the capac-itor falls to zero, and the capacitor starts to charge in the opposite direction (Fig-ure 10-6c). The current flow is in the same direction as in Figure 10-6b, but the voltage polarity across the capacitor has changed.

6. At time position 6, current flow is dropping off. The voltage across the capacitor approaches maximum negative.

7. At time position 7, current flow is back to 0. The voltage is at maximum negative.

8. At time position 8, current flow has again changed direction. The voltage across the capacitor starts to decrease (Figure 10-6d).

9. At time position 9, current flow is maximum. The voltage across the capacitor rapidly passes through 0, and the capacitor starts to charge. This completes one cycle of ac in a span of $\frac{1}{60}$ second. This means that there are 60 repeats each second in the graph in Figure 10-5.

CAPACITOR CONSTRUCTION

For practical reasons, capacitors use thin metal sheets, usually aluminum, for the plates. Wax paper, oil-saturated paper, or a chemical electrolyte is used as the high-resistance material between the plates. For air-conditioning and refrigeration service, the two common types of capacitors are the oil-filled and the electrolytic capacitors (Figures 10-7a and b).

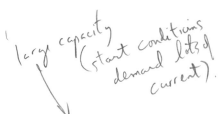
A.C. capacitors current charges/discharges all the time --- heat must be dissipated

FIGURE 10-7(a) Oil-filled capacitors (run capacitors). (Courtesy of BET, Inc.)

large capacity (start conditions demand lots of current).

FIGURE 10-7(b) Electrolytic capacitors (start capacitors). (Courtesy of BET, Inc.)

↳ A surge of D.C. current

Oil-filled capacitors are used in series with the start winding of ac motors in order to increase the torque of the motor. Oil-filled capacitors are also used when the application requires the capacitor to continuously be in the circuit.

Electrolytic capacitors are used in the start circuits of high-horsepower ac motors. An electrolytic capacitor provides a high capacitance in a small space. Electrolytic capacitors are usually connected for intermittent use.

CAPACITOR RATINGS

Capacitors are usually rated by capacity and a maximum voltage rating. This capacity unit is the *farad*, which is a very large unit. Capacitors are usually rated in microfarads (one-millionth of a farad), which is a practical unit for most applications. Some capacitors are marked 10 MFD, 10 mfd, or 10 MF, each of which means 10 microfarads. The common abbreviation for microfarad is μF.

Whenever a capacitor is replaced in an electrical device, the capacitance values should be matched as closely as possible. The voltage rating of a capacitor is also an important consideration. The voltage rating of the replacement capacitor should equal or exceed that of the original capacitor.

Standard oil-filled capacitors for continuous use are usually available in capacitance values up to 50 μF, with voltage ratings up to 500 V. Electrolytic capacitors for intermittent use are usually available in capacitance values of 70 to 800 μF at 125 V and up to 200 μF at 440 V.

The most common purpose for capacitors in air-conditioning and refrigeration circuits is to provide starting torque and to maintain running torque in ac motors. Further treatment of capacitive circuits will be given in Unit 13.

INDUCTANCE

Laws covering inductance in electrical circuits have been presented in earlier units. One of the most important is that whenever a conductor (wire) is cut by a magnetic field, a voltage is induced in the wire.

Consider a coil of wire with an ac voltage applied to it. The current flow through the coil will be continuously variable: starting at zero, rising to a maximum, returning to zero, then rising to a maximum in the opposite direction before returning to zero again (Figure 10-8).

When a continuously varying current (ac) flows through a coil of wire, it produces a continuously varying magnetic field about the coil. This varying magnetic field cuts through the coil and produces a voltage in the coil. The voltage that is

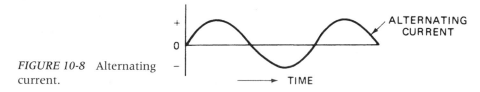

FIGURE 10-8 Alternating current.

induced is in direct opposition to the supplied voltage. The induced voltage is called *back electromotive force,* back EMF, or BEMF.

In Figure 10-9, a coil of wire is shown with a voltage of 100 volts at 60 hertz ac supplied to it. The ammeter indicates 1 ampere of current flow. The dc resistance of the coil is given as 1 ohm. According to Ohm's law, only 1 volt is needed to force 1 ampere through the 1-ohm resistance of the coil. The remaining 99 volts overcome the 99 volts of back EMF developed in the coil. The back EMF of the coil has a greater effect in the control of current than in the resistance of the coil.

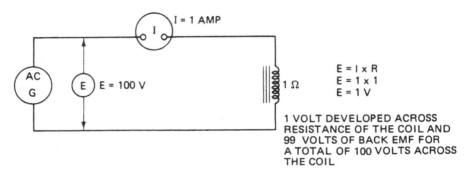

FIGURE 10-9   Back EMF generated in a coil.

INDUCTIVE EFFECTS

Coils of wire are used in many applications in air-conditioning and refrigeration electrical systems. The most common uses are in motors, transformers, relays, and solenoids. In each type of device, back EMF is a major factor in the control of current.

CURRENT FLOW IN INDUCTIVE CIRCUITS

The ac current flow through a circuit containing a coil varies (Figure 10-10). The magnetic field that develops around a coil carrying current is proportional to the current. As the current changes, so does the magnetic field. Around point A in Figure 10-10, the current is changing very little. The current rises slowly to a maximum and then decreases to a low level. At the maximum value of current, point A, the current is not increasing or decreasing; the rate of change is zero. At this same point, the change in the magnetic field is zero. Without a changing magnetic field, the induced voltage (back EMF) is zero.

The maximum voltage is induced in a coil when the rate of change of the magnetic field is greatest. In Figure 10-10, the maximum rate of change of current is around point B when the current passes through zero. The maximum rate of change in the magnetic field is at the same time. The maximum voltage (back EMF) also occurs at this time.

A plot of this relationship is shown in Figure 10-11. This is a graph of the current through and voltage across a pure inductance coil.

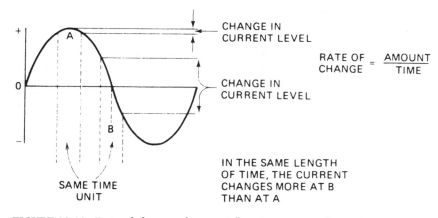

FIGURE 10-10 Rate of change of current flow is greatest when current passes through zero.

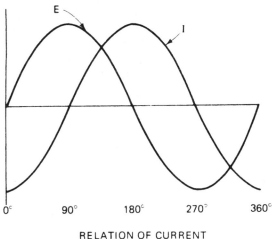

FIGURE 10-11 Current lags voltage by 90° in a pure inductance.

1. The voltage is maximum as the current passes through zero.

2. The voltage is zero as the current reaches maximum.

It is not necessary that current and voltage rise and fall together in ac circuits. Actually, it is more common for them not to do so, since so many electrical circuits contain coils. Further treatment of coils and the angular relationship between current and voltage will be given in Unit 13.

SUMMARY

- In a capacitor, energy is stored in the form of an electrostatic field.

- In an inductive circuit, energy is stored in the magnetic field.

- Capacitance in a circuit opposes a change in voltage.
- Inductance in a circuit opposes a change in current.
- If a circuit contains inductance and or capacitance, the current and voltage in the circuit will likely be out of phase with each other.

PRACTICAL EXPERIENCE

Required equipment   Analog ohmmeter, various capacitors, special bleed resistor (see Figure 10-12).

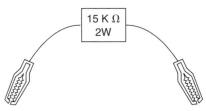

FIGURE 10-12   15KΩ 2-watt bleed resistor with wire and insulated clip leads attached.

Procedure

1. Using the bleed resistor with clip leads, carefully connect the resistor across the capacitor. Make connection to one terminal of the capacitor at a time.
2. List the sizes in MF of each capacitor to be tested.
    a. _____MF
    b. _____MF
    c. _____MF
    d. _____MF
    e. _____MF

    NOTE: A standard speed count sequence will be used to estimate the time it takes for each capacitor to charge.

3. Place the ohmmeter selector switch on the highest $R \times$ position (probably $R \times 10,000$).
4. In turn, connect the ohmmeter across each capacitor. Start counting when the connector is made.
5. Record the count (turn) it takes for each capacitor to charge to the meter's internal battery voltage.

    NOTE: Use a fixed point-size 1-meg ohm as the measure with the scale of $R \times 10,000$. A meter reading of 100 is 1 meg ohm.

	Size	*Count*	
1._____		MF_____	
2._____		MF_____	
3._____		MF_____	
4._____		MF_____	

Question: Does it take a longer time to charge large MF values or small MF values?

Conclusions

1. Capacitance and inductance play many important roles in air-conditioning systems.

2. In air-conditioning circuits, the current and voltage are nearly in phase with each other. → ?

3. Back EMF (BEMF) is generated in coils.

REVIEW QUESTIONS

1. Name the two common types of capacitors used in air-conditioning and refrigeration systems.

2. A capacitor is made up of two metal plates separated by some _____ material.

3. Which type of capacitor is generally used intermittently in compressor motor-starting circuits?

4. Two important capacitor ratings are _____ and _____.

5. What is the relationship between the voltage induced in a coil carrying current and the voltage that produced the current?

6. With ac current, the rate of change is greatest when the current passes through _____.

7. When the ac voltage applied to a capacitor is maximum, the current flow into the capacitor is _____.

8. The (electrolytic, oil-filled) _____ capacitor usually has the higher microfarad value in a motor circuit.

9. Refer to Figure 10-10 on current flow in an inductive circuit. The greater induced voltage across the coil takes place at point (*A* or *B*) _____.

10. Refer to Figure 10-11. In this voltage and current relationship, the current is shown (leading, lagging) _____ the voltage by _____ degrees.

Unit 11

Electrical Power and Energy

OBJECTIVES

Upon completion and review of this unit, you will be able to
- Calculate electrical power using voltage, current, and resistance.
- Calculate electrical energy using electrical power and time.
- Control electrical power using relays and contactors.

Power is a measure of the rate of doing work. Mechanical power is measured in horsepower (hp), and electrical power is measured in watts (w). The unit of electrical power is named for James Watt (1736–1819), a Scottish engineer who developed the modern condensing steam engine. In the electrical system, power is equal to voltage times current.

$$W = I \times E$$

An electric heat strip unit of an air-conditioning system may have a hot resistance of about 6 ohms (Figure 11-1). (The cold resistance of this unit would be only slightly lower.) When in use, 9600 watts of power is produced in the unit. The unit is designed for 240 volts at 40 amperes.

$$I = \frac{E}{R}$$

$$I = \frac{240}{6}$$

$$I = 40 \text{ amps}$$

$$W = I \times E$$

$$W = 40 \times 240$$

$$W = 9600 \text{ watts}$$

This may be seen written on nameplates as 9.6 kw, which is 9.6 thousand watts. The k stands for one thousand.

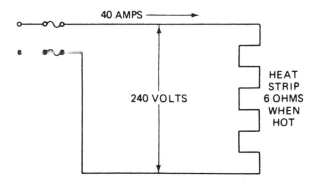

40 AMPS ⟶

240 VOLTS

HEAT STRIP 6 OHMS WHEN HOT

FIGURE 11-1 Heat strip resistor.

A good example of a power-consuming device is an electric bread toaster. In Figure 11-2, an electric toaster is shown plugged into a 120-volt electrical source. Ten amperes of current is being drawn from the power source. The power developed in the toaster may be determined by the following formula: watts = amperes × volts.

$$W = I \times E$$

$$W = 10 \times 120$$

$$W = 1200 \text{ watts} \quad \text{or} \quad 1.2 \text{ kw}$$

Another formula is sometimes used to determine power in electrical circuits.

$$W = I^2R$$

As a statement, the formula is that power (W) is equal to amperes (I) squared times resistance (R). Another way of stating the formula is that power (W) is equal to amperes (I) times amperes (I) times resistance (R).

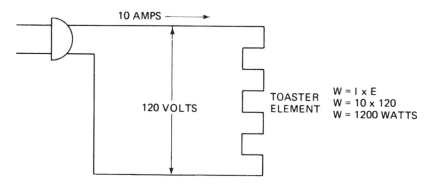

10 AMPS ⟶

120 VOLTS

TOASTER ELEMENT

W = I x E
W = 10 x 120
W = 1200 WATTS

FIGURE 11-2 Standard toaster.

Since in Ohm's law $E = I \times R$, the factors $I \times R$ can be substituted for E in the original power formula.

$$W = I \times E$$

Substituting $(I \times R)$ for E,

$$W = I \times (I \times R)$$

$$W = I \times I \times R$$

$$W = I^2R$$

In the toaster example, the resistance of the toaster heating element, according to Ohm's law, is

$$R = \frac{E}{I}$$

$$R = \frac{120}{10}$$

$$R = 12 \text{ ohms}$$

When this factor is used in the formula $W = I^2R$,

$$W = 10 \times 10 \times 12$$

$$W = 100 \times 12$$

$$W = 1200 \text{ watts}$$

This same answer was obtained using the formula $W = I \times E$.

ELECTRIC MOTORS

Electric power is required to operate electric motors. In a motor, the desired mechanical output is a rotation of the motor shaft. Some electrical power is continually lost in a motor because of the heat produced. A motor is, therefore, not 100% efficient.

Consider an electric motor used to drive a refrigeration compressor (Figure 11-3). The compressor is rated as 1 horsepower (hp). The motor, also rated as 1 hp, is delivering 1-hp mechanical power to the compressor. With 4 amperes of

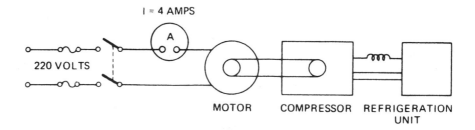

FIGURE 11-3 Motor rated in output horsepower.

current being drawn from the power line (at 220 volts), 880 watts of electrical power is being delivered to the motor.

$$W = I \times E$$

$$W = 4 \times 220$$

$$W = 880 \text{ watts}$$

If the motor is operating at about 85% efficiency, about 134 watts is converted into heat in the motor. The remaining 746 watts is converted into mechanical rotating power at the motor shaft (output).

The conversion between horsepower and watts is 746 watts equals 1 horsepower.

Therefore, with 746 watts being converted to rotating power, there is 1 hp being delivered to the motor shaft. Motors are rated in shaft horsepower, that is, output, not power input. As an aid in remembering the hp/watt conversion factor, remember: Columbus discovered America in 1492; $1492 \div 2 = 746$; 746 watts = 1 horsepower.

ELECTRIC ENERGY

Electric power was defined earlier as a measure of the *rate* of doing work. Electric energy relates to the total amount of work done. Stated a different way, the rate of doing work (power) is multiplied by time.

$$\text{Energy} = \text{rate} \times \text{time}$$

The power company charges its customers for the total amount of electrical energy it supplies. Since 1 watt-hour (wh) is a very small unit of energy, the power company charges per kilowatt hour. A kilowatt hour (kwh) is equal to 1000 watts for 1 hour, 500 watts for 2 hours, 2000 watts for $\frac{1}{2}$ hour, and so on.

$$1000 \text{ watts} \times 1 \text{ h} = 1000 \text{ wh (1 kWh)}$$

$$500 \text{ watts} \times 2 \text{ h} = 1000 \text{ wh (1 kWh)}$$

$$2000 \text{ watts} \times 0.5 \text{ h} = 1000 \text{ wh (1 kWh)}$$

where kilo (k) stands for one thousand. The power company bases its charge for providing electrical energy on its cost of production plus a small profit. The cost varies by geographic location and is obtained only from your local company. The following example, based on $0.06 per kwh, illustrates how the cost of operation may be determined when power consumption is known.

EXAMPLE 1 _____

An electric power company charges its customers 6 cents for 1 kilowatt hour (kw/h) of energy. How much does it cost the customer to heat for 8 hours if the heat is obtained from a heat strip of 22 ohms connected to a 220-volt source (Figure 11-4)?

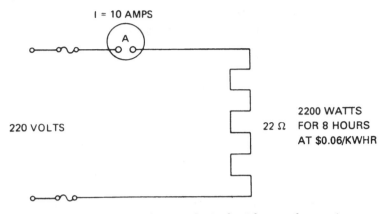

FIGURE 11-4   Heat strip costs $1.06 for 8 hours of operation.

Solution   Solve for current:

$$I = \frac{E}{R}$$

$$I = \frac{220}{22}$$

$$I = 10 \text{ amperes}$$

Solve for power:

$$W = I \times E$$

$$W = 10 \times 220$$

$$W = 2200 \text{ watts}$$

Solve for energy:

$$W/h = W \times time$$

$$W/h = 2200 \times 8$$

$$W/h = 17,600 \text{ watt-hour or } 17.6 \text{ kWh}$$

Solve for cost 6 cents per kWh:

$$Cost = kw \times 0.06$$

$$Cost = 17.6 \times 0.06$$

$$Cost = \$1.06$$

For the power company to keep track accurately of the total amount of electrical energy being supplied to a customer, a watt-hour meter is usually installed where the electric service is delivered to a building. A watt-hour meter is an electromechanical device that rotates dials, recording the energy delivered to the customer (Figure 11-5). This meter is read monthly by the power company, and the customer is billed for the energy used.

FIGURE 11-5 Watt-hour meter. (Courtesy of BET, Inc.)

CONTROL OF ELECTRIC POWER

The control of electric power is a problem that must be dealt with every day. Electric power must be turned off and on either manually or automatically as needed for lights, motors, heaters, and other power-consuming devices.

In low-power units, such as lamps, electric drills, and electric toasters, simple switches may be used to provide the required control. In the larger power-consuming devices, such as air-conditioning compressor units or electric-heat strip units, high power (usually at higher voltages) must be controlled. In these types of units, power must be controlled at a point remote from where it is being consumed. For example, an air-conditioning compressor (often outside the house) is controlled by a thermostat in the living area inside the house. A low-voltage signal (usually 24 volts) is fed to a relay (an electrically operated switch) in the air-conditioning unit. The relay, when energized, supplies high voltage (usually 220 volts) at high current levels to the compressor motor.

Relays are electromechanical devices used as controls for electric power. A relay has a movable metal part called the armature. An electrical contact or contacts may be connected to the armature; when the armature moves, the contacts also move. Depending on the application, this action may open some electric circuits while closing others.

In Figure 11-6, the relay is in the de-energized (no coil power applied) position. When no power is supplied to the relay coil, the spring pulls the armature down at the right end. The armature pivots at the point, moving the left end of the armature up. An electric circuit is complete from the normally closed contact through the movable contact to the connecting wire to the terminal output.

In Figure 11-7, the same relay is shown, but electric power has been applied to the relay coil. The magnetic field developed around the coil attracts the iron in the armature. The armature moves down, making electrical contact between the normally open contact and movable contact. As long as power is supplied to the coil, the relay remains energized. When power is removed from the coil, the spring

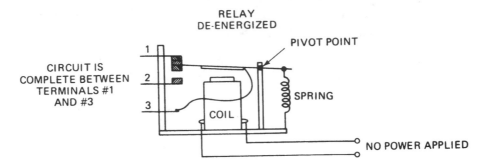

FIGURE 11-6   Relay armature position, de-energized.

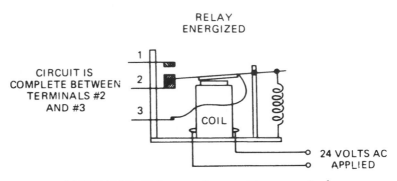

FIGURE 11-7   Relay armature position, energized.

pulls the armature to its original position, returning the connection to the normally closed contacts.

Relays come in varied shapes and sizes and are designed for special purposes. Sometimes the name of the device, because of its application, is changed, such as to *contactor*. The mechanical construction of a contactor (Figure 11-8) is somewhat

FIGURE 11-8   Contactor. (Courtesy of BET, Inc.)

different from that of a relay, but it operates on the same principle. Low power supplied to the coil of the contactor produces a magnetic field that attracts an armature. When the armature moves, it closes electric contacts that provide for the control of high-power circuits.

More will be said about relays and contactors in Unit 16.

MEASURING ELECTRIC POWER

One method of determining electric power in a circuit is to measure the voltage across the circuit and the current through the circuit. The power developed is found by the formula

$$W = I \times E$$

In Figure 11-9, the circuit of an electric heat strip is shown. The voltage across the heat strip is 220 volts as indicated by the voltmeter. The ammeter indicates 43.6 amperes.

$$W = I \times E$$

$$W = 43.6 \times 220$$

$$W = 9600 \text{ watts}$$

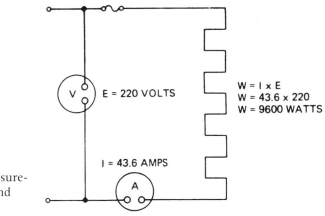

FIGURE 11-9   Power measurement using a voltmeter and ammeter.

Another method of measuring power is to use a wattmeter. A *wattmeter* is a device that uses both the current in the circuit and the voltage across the circuit to provide an indication of the power developed in the circuit. The heat strip of Figure 11-9 could have a wattmeter connected rather than a voltmeter and an ammeter. The power developed would be read directly from the wattmeter, as shown in Figure 11-10.

It is always better to use a wattmeter for determining power, if one is available. Since only one meter reading is needed, the chance of error is lessened. In ac circuits, it is always better to use a wattmeter for power measurement because of other factors having effects on ac. Further information on power in ac circuits will be covered in Unit 13.

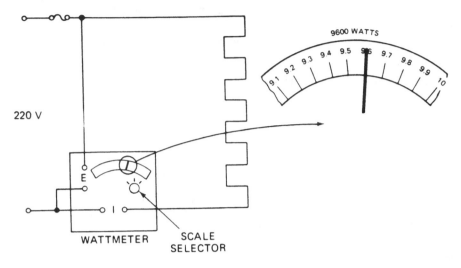

FIGURE 11-10   Power measurement using a wattmeter.

SUMMARY

- Electric power may be calculated using formulas:

$$W = I \times E$$

$$W = I \times R$$

$$W = E \times R$$

- Wattmeters indicate electrical energy through the use of current measurement and voltage measurement in the same instrument.
- Electrical energy is electrical power times time. It indicates the total work done.

PRACTICAL EXPERIENCE

Required equipment   Heat strip, ac voltmeter, ohmmeter, and clamp-on ammeter.

Procedure

1. Observe the nameplate specifications of the heat strip.
2. Using the ohmmeter measure the cold resistance of the heat strip._____
3. Connect the heat strip to the proper voltage as indicated on the nameplate.
4. Measure the voltage at the heat strip terminals._____
5. Use the clamp-on ammeter to measure the current flow through the strip.
6. Calculate the power developed in the heat strip.

$$W = I \times E = \underline{\hspace{2cm}} \text{ watts}$$

7. Calculate the hot resistance of the heat strip.

$$R = \frac{E}{I} = \underline{\hspace{2cm}} \text{ ohms}$$

8. Is the hot resistance higher or lower than the cold resistance of the strip?

REVIEW QUESTIONS

1. The formula for power using voltage and current is $W = \underline{\hspace{1.5cm}}$.

2. The formula for power using current and resistance is $W = \underline{\hspace{1.5cm}}$.

3. A strip heater draws 30 amperes of current at 220 volts. How much power is developed in the unit?

4. If a strip heater of 8600 watts is operated for 100 hours, how much electrical energy will be converted to heat?

5. A power company is selling electric energy at $0.04 per kWh. How much will it cost to operate a 10,000-watt heater for 60 hours?

6. The movable part of a relay is called an $\underline{\hspace{1.5cm}}$.

7. A contactor is a special-purpose $\underline{\hspace{1.5cm}}$.

8. In Figure 11-11, if the power developed in the heater is 12,000 watts and the supply voltage is 220 volts, what is the current flow through the heater?

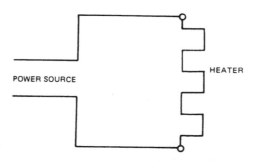

POWER SOURCE   HEATER

FIGURE 11-11   Strip heater.

9. In Figure 11-11, the power developed in the heater is 8000 watts. The power source is 220 volts. What is the resistance of the heater?

10. The circuit of Figure 11-11 is to supply 18,000 watts of heat. The power source is 220 volts. What is the resistance value of the heater?

Unit **12**

Transformers

OBJECTIVE

Upon completion and review of this unit, you will have learned
- Transformer theory; the relation between transformer turns; and ratio, voltage, and current.
- Of the power losses in transformers.
- Of delta and wye connections of transformers.
- How transformers are rated.

Transformers are nonmechanical electric devices used in many applications for the control of electrical power. High voltages used on transmission lines for electrical power transfers are dropped (reduced) with the use of transformers. Voltages, at low levels, used in air-conditioning and refrigeration control systems are obtained through the use of transformers.

For example, the transmission cable may be 2400 volts, while utility service requirements are 240 volts. A transformer is used to reduce the 2400-volt transmission to a usable 240 volts. This 240 volts supplies power to the air-conditioning system's high-voltage electrical components. Another transformer within the electrical system steps this 240 volts down to 24 volts for the low-voltage control system.

VOLTS PER TURN

In Unit 8, the back EMF developed in a coil of wire carrying an ac current was discussed. The relationship was given that showed that the inductance of the coil has the major effect in the control of current. The varying magnetic field around a coil carrying an ac current cuts through the coil and produces a back EMF that limits the total current. The small current that flows through the coil is called the *magnetizing current.*

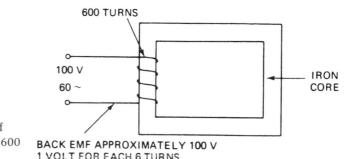

600 TURNS

100 V
60 ~

IRON
CORE

BACK EMF APPROXIMATELY 100 V
1 VOLT FOR EACH 6 TURNS

FIGURE 12-1 Back EMF of 0.166 volts per turn in the 600 turn coil.

Figure 12-1 shows a coil of wire wound on an iron core. The coil of wire has 600 turns in it, connected in series. The back EMF developed in the coil is very close to 100 volts. Another way of stating the situation is that for every six turns of wire on the coil approximately 1 volt is developed.

A second coil of wire is wound on the same iron core (Figure 12-2). The second coil is called the *secondary winding;* the first coil is called the *primary winding.* The same magnetic field that cut through the primary winding, developing back EMF, cuts through the secondary winding. A voltage is developed in the secondary winding at 1 volt per 6 turns. Since there are 60 turns in the secondary, a total of 10 volts is developed across the secondary winding (60 ÷ 6 = 10).

There is a voltage relationship between the primary and secondary windings of a transformer. Simply stated, the turns ratio is equal to the voltage ratio. The transformer shown in Figure 12-2 has a turns ratio of 600 to 60 or 10 to 1; the voltage ratio is also 10 to 1.

Figure 12-3 shows the same transformer with a load resistance of 1 ohm connected across the secondary. The 10-volt output across the secondary winding will be impressed across the 1-ohm resistor after the switch is closed. According to Ohm's law,

$$I = \frac{E}{R}$$

$$I = \frac{10}{1}$$

$$I = 10 \text{ amperes}$$

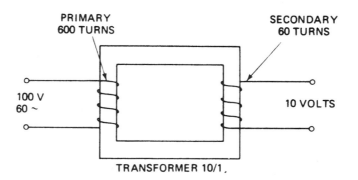

PRIMARY
600 TURNS

SECONDARY
60 TURNS

100 V
60 ~

10 VOLTS

TRANSFORMER 10/1.

FIGURE 12-2 Voltage is developed in the secondary at 0.166 volts per turn.

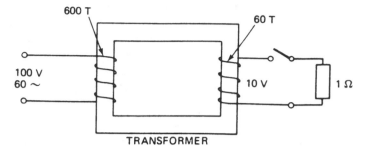

FIGURE 12-3 Current flow of 10 amperes developed 100 watts in the load resistor.

The 10 amperes of current flow through the 1-ohm resistor. Electrical power is supplied by the secondary of the transformer. This power must be coming through the primary of the transformer from the 100-volt, 60-hertz power source. The power developed in the secondary circuit is

$$W = I \times E$$

$$W = 10 \times 10$$

$$W = 100 \text{ watts}$$

When the primary of a transformer is connected to a source voltage, a small current, called the magnetizing current, flows through the primary. The magnetic field set up by this magnetizing current is indicated in Figure 12-4 as ①.

When a load is connected across the secondary, current flows in the secondary. This current produces a magnetic field in the core of the transformer in di-

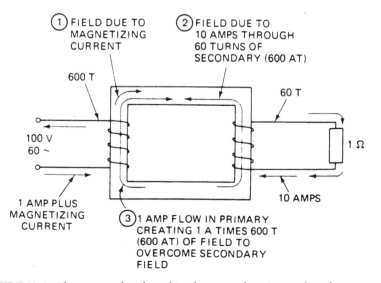

FIGURE 12-4 The power developed in the secondary is equal to the power supplied to the primary.

rect opposition to the field set up by the magnetizing current. In Figure 12-4, this magnetic field is indicated as ②. The field strength is based on 600 ampere-turns and tends to reduce the back EMF of the primary. The decreased back EMF allows for an increase in current in the primary. → *Secondary induced magnetic field is opposite to primary induced magnetic field.*

A current flow producing a magnetic field based on 600 ampere-turns is required in the primary to overcome the 600 ampere-turns in the secondary. With 600 turns in the primary, 1 ampere of current is needed to produce this field. The field is shown as 3 in Figure 12-4.

The relationship between the currents in the primary and the secondary can now be determined. The currents are inversely proportional to the turns ratio. For example, in Figure 12-3

1. The turns ratio is 600/60 or 10/1.

2. The voltage ratio is 600/60 or 10/1.

3. The current ratio is 60/600 or 1/10.

The power developed in the secondary has been shown to be 100 watts. The power developed in the primary is

$$W = I \times E$$
$$W = 1 \times 100$$ *to be precise, 100 VA.*
$$W = 100 \text{ watts}$$

Actually, the current flow in the primary is slightly more than 1 ampere since the magnetizing current also flows in the primary. The transformer is not 100% efficient; there is some power loss.

POWER LOSS IN TRANSFORMERS

There are three main reasons for power loss in a transformer:

1. Resistance of the windings

2. Hysteresis

3. Eddy currents

Resistance of the windings. The power loss in the resistance of the windings needs no further explanation. Whenever current flows through resistance, heat is developed and power is lost. This power loss is known as the I^2R loss.

Hysteresis. When alternating current flows in a coil, the magnetic field about the coil changes direction with the current. If the coil has an iron core, as in a transformer, the direction of the magnetism of the iron core must also change with the current: Each magnetic molecule of the iron core must change direction with each alternation of the current. This continuous motion of the magnetic molecules produces heat due to friction, which is a power loss in the transformer.

Eddy currents. Eddy currents are set up in the iron core of a transformer. The core of the transformer is a metal, and it acts the same as a secondary winding. A voltage is induced in the core, and current flows in the core. Eddy currents flow throughout the core. If solid iron cores were used in transformers, the losses due to eddy currents would be quite high.

To overcome the high-eddy current losses, the cores of transformers and most other magnetic devices using alternating current are laminated. *Laminations* are thin, sheet-metal parts that may be mechanically fastened together to produce the required core. Through the use of thin sheets of metal, the eddy current is confined to the individual lamination. Because the resistance path for eddy currents is increased, the total eddy current draw is decreased. Laminated iron cores are found in most ac devices, such as transformers, motors, solenoids, relays, and contactors.

POWER TRANSFORMERS

Power transformers are used throughout the world to step up voltages for transmission and distribution. They are also used to step down voltages for use at the destination point.

For example, a power company may generate electric power at a voltage level of 13,800 volts at the generating station. The main factor in determining the voltage level at the generator is the insulation problem related to operating generators at higher voltages. Before the electrical power is distributed on transmission lines across country to the individual localities, the voltage level is stepped up, using transformers. A standard transmission line voltage level is 69,000 volts. Substations (transformers) are used to reduce the voltage to safer levels (about 2400 volts) for distribution around populated areas. Finally, at the end point of use, the voltage is further reduced by a transformer to the level required by the consumer, usually 220/110 volts. Figure 12-5 shows an example of a full system from generation to final use of electrical power by the residential customer. The voltage levels shown are representative of industry standards.

Power transformers at substations are usually large, bulky items (Figure 12-6). High-voltage insulators of ceramic material are usually visible at the terminals of the transformers. Power transformers used to step voltage down from a final substation (2400 volts) to the voltage needed in residential use (220/110 volts) are located on power-line poles (Figure 12-7). A single transformer will usually supply several homes. The final step-down transformer is center tapped and therefore can provide the required 220 volts for power circuits and also the 110 volts required for lighting and small appliances.

Transmission Losses

Voltages are stepped up whenever power is to be distributed over long distances, in order to decrease power losses. For example, consider a manufacturing plant located approximately 10 miles from a power-generating station (Figure 12-8). The manufacturing plant requires 100,000 watts of electrical power. The generating station is producing power at a 10,000-volt level. A current of 10 amperes is needed at the 10,000-volt level to provide the 100,000 watts.

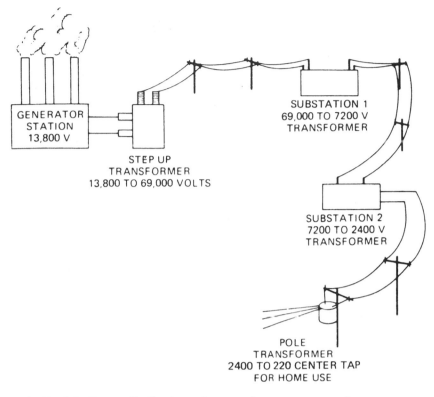

FIGURE 12-5   Power distribution using transformers to step voltage up and down.

FIGURE 12-6   Transformer substation. (Courtesy of Florida Power & Light)

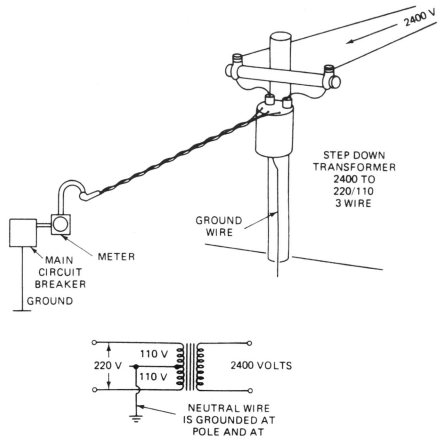

2400 V

STEP DOWN
TRANSFORMER
2400 TO
220/110
3 WIRE

GROUND
WIRE

MAIN
CIRCUIT
BREAKER

METER

GROUND

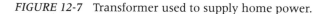

220 V

110 V

110 V

2400 VOLTS

NEUTRAL WIRE
IS GROUNDED AT
POLE AND AT
SERVICE ENTRANCE

FIGURE 12-7   Transformer used to supply home power.

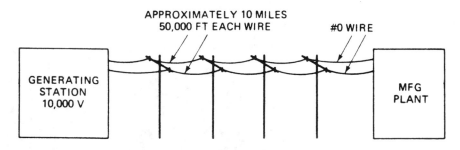

APPROXIMATELY 10 MILES
50,000 FT EACH WIRE

#0 WIRE

GENERATING
STATION
10,000 V

MFG
PLANT

FIGURE 12-8   Power distribution at 10,000 volts (high voltage and power loss
in lines).

Two wires are required to get the electrical power to the manufacturing plant; each wire is 50,000 feet long. If size 0 wire is used, the total resistance of the two wires will be about 10 ohms, because size 0 wire has 0.1 ohm per 1000 feet. According to Ohm's law, the voltage developed on the resistance of the transmission line would be 100 volts.

$$E = I \times R$$

$$E = 10 \times 10$$

$$E = 100 \text{ volts}$$

This would leave only 9900 volts to be delivered to the manufacturing plant. The power company would continually waste the power developed in the resistance of the transmission lines.

$$W = I \times E$$

$$W = 10 \times 100$$

$$W = 1000 \text{ watts lost}$$

The losses can be reduced and additional savings can be realized by employing transformers, thus making feasible the use of smaller wire in the transmission lines. In Figure 12-9, the power station and the manufacturing plant are the same. A step-up transformer has been added at the generator station, and a step-down transformer has been added at the manufacturing plant. The transmission lines have been changed to size 3 wire, which has about 0.2-ohm resistance per 1000 feet. The resistance of the total transmission line, two lengths of 50,000 feet each, is about 20 ohms. With the step-up of the voltage at the power plant from 10,000 to 100,000 volts, the current is stepped down from 10 to 1 ampere. The power loss in the lines has been reduced considerably.

The voltage developed in the resistance of the transmission lines is

$$E = I \times R$$

$$E = 1 \times 20$$

$$E = 20 \text{ watts} \text{volts}$$

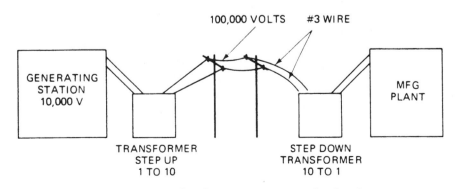

FIGURE 12-9 Power distribution at 100,000 volts (low losses).

The power loss is

$$W = I \times E$$

$$W = 1 \times 20$$

$$W = 20 \text{ watts}$$

The voltage at the primary of the step-down transformer is

$$100,000 - 20 = 99,980 \text{ volts}$$

At the secondary, the voltage is 9998 volts, which is much closer to the 10,000 volts being delivered by the generating station.

In actual practice, the voltage required at a manufacturing plant is seldom that generated at a power station. The examples given of using transformers to step up and step down voltages to save power are similar to the methods used in actual situations to adjust voltage at the receiving user.

POWER DISTRIBUTION: DELTA AND WYE

Almost all electrical power is generated using three-phase alternators. Power transformers are used to step the voltages up or down as required by the distribution system. The turns ratio of the transformers is not the only factor to be considered when three-phase power is to be transformed. The transformer connections, delta or wye, must be taken into account.

Three-Phase Delta-Wye Combinations

The primaries and secondaries of transformers used in three-phase systems may be connected in delta or wye configuration. The output voltage is determined not only by the input voltage and the turns ratio. The delta or wye connection of the primaries and secondaries also affects the output voltage. Consider the four possible connections of the three transformers in Figure 12-10. The transformers are step down, with a 5/1 turns ratio. The input voltage is 1000 volts.

In Figure 12-10a, the transformers are connected delta to delta. The primary coil voltage is equal to the line voltage. The secondary line voltage is equal to the secondary coil voltage.

In Figure 12-10b, the transformers are connected wye to wye. The primary coil voltage is equal to the line voltage divided by the square root of 3 (1.732). There are 577 volts across each primary coil. The secondary coil voltage is reduced by a factor of 5 due to the turns ratio. There are 115 volts across each secondary coil. The voltage from any line to ground is 115 volts. The voltage between any two lines is the coil voltage times the square root of 3, or 200 volts (115 V $\times$ 1.732 = 200 V).

In Figure 12-10c, the transformer primaries are delta connected. The secondaries are wye connected. There are 1000 volts across each primary coil. The step-down turns ratio provides 200 volts across each secondary coil. Since the secondaries are wye connected, the voltage line to ground is the voltage of the coil,

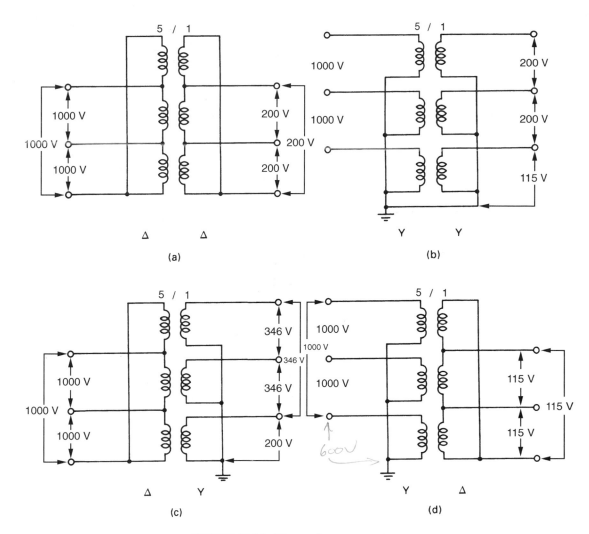

FIGURE 12-10 Three-phase connectors.

or 200 volts. The voltage between any two lines is the coil voltage times the square root of 3, or 346 volts (200 V × 1.732 = 346 V).

In Figure 12-10d, the primaries are connected in wye. The transformer primary coil voltage is equal to the line voltage divided by the square root of 3, or 577 volts. The 5/1 turns ratio provides a secondary coil voltage of 115 volts. The secondaries are delta connected. The line voltage is equal to the coil voltage, 115 volts.

With the delta connection the line voltage equals the coil voltage.

With the delta connection the coil voltage equals the line voltage.

With the wye connection the line voltage equals the coil voltage times 1.732.

With the wye connection the coil voltage equals the line voltage divided by 1.732.

With the delta connection the line current equals the coil current times 1.732.

With the delta connection the coil current equals the line current divided by 1.732.

With wye connection the line current equals the coil current.

With wye connection the coil current equals the line current.

Standard Delta-Connected Secondaries

Figure 12-11 shows the secondary connection in a standard delta system. The voltage between any two lines is 240 volts. The voltage between line A and neutral or line C and neutral is 120 volts. The voltage between line B and neutral is approximately 208 volts.

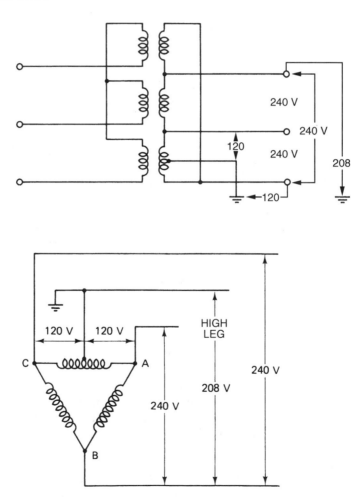

FIGURE 12-11   Delta secondary with 208-V high leg.

In large commercial and office installations, the 240 volts could be used to operate power equipment such as air conditioners. The 120 volts could be used to supply the outlets. The phase *B* to ground voltage is 208 volts. This phase is often called the *high leg* or *wild leg*. Normally, this phase is identified in panel boxes by orange tape or paint. Phase *B* is normally placed in the center lug of a panel box. Special care must be taken not to use this phase to ground for 120 volts. The higher voltage (208 V) could damage 120-volt equipment.

Open Delta

Three-phase power is sometimes provided using the open-delta system. An advantage of the open-delta system is that only two transformers are required for the three-phase system. The open-delta connection is shown in Figure 12-12. Note that the high leg or wild leg, phase *B*, is also present in the open-delta system.

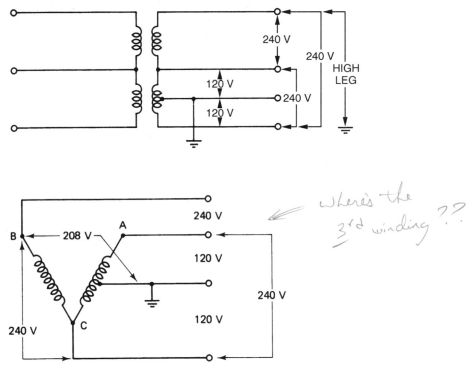

FIGURE 12-12   Open delta. (Note high leg.)

SIGNAL TRANSFORMERS

Transformers are used in almost all air-conditioning units to provide safe, low-level ac voltage for control purposes. The input voltages for air-conditioning systems may be 110, 220, or 440 volts; 220 volts is the most common input for residential control systems. A step-down transformer within the air-conditioning unit provides 24 volts ac for use in the control circuit. The control circuit is used to energize high-voltage circuits for the control of motors and other devices.

An example of a low-voltage control circuit of an air-conditioning system using a 24-volt step-down transformer is shown in Figure 12-13. A voltage of 220 volts is fed through the disconnect switch to the primary of the low-voltage transformer and to the top of the two power contactors. With the thermostat switches (covered in Unit 16) as shown, the 24 volts from the secondary of the low-voltage transformer is fed to the coil of the compressor contactor and to the coil of the inside fan motor; both contactors energize. The 220 volts is then fed through both contactors to the motors.

After the room cools, the thermostat contacts open at a preselected temperature. This removes the 24 volts from the contactor coils; the contactors de-energize. The contacts then open and 220-volt power is removed from the motors. The motors will not run until the room temperature rises and the switches (contacts) in the thermostat again close.

Through the use of a low-voltage transformer, a safe system of signal transfer is accomplished. The individual making adjustments at the thermostat is not in close proximity to dangerous high-voltage power circuits.

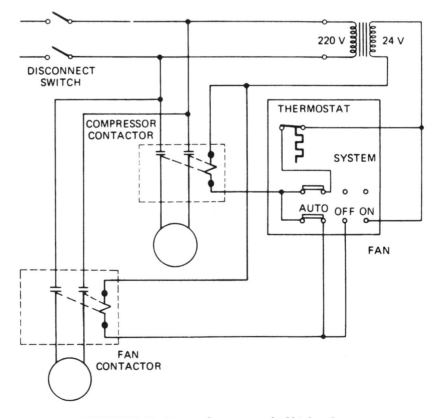

FIGURE 12-13   Low-voltage control of high voltage.

Another important factor in the use of low voltage in air-conditioning control circuits is the cost of installation of control-circuit wiring. If 110 volts were used in the control circuit, wiring between the thermostat and the evaporator/condenser would have to be treated the same as the power circuits for lighting or outlets. If the local building code called for the use of conduits in all line-voltage circuits, then a conduit would have to be installed between the evaporator/condenser unit and the thermostat. Installation costs could vary greatly, depending on the distance between the thermostat and the condenser.

The same installation using a low-voltage control system would be less expensive. Control circuit wiring between the evaporator/condenser and thermostat would consist of a standard signal cable that met building code requirements. If the conduit is not needed, the cost decreases dramatically.

TRANSFORMER RATINGS

Transformers are usually rated with the primary and secondary voltage and the volt-ampere (VA) design of the unit. An example of this is a control transformer. This transformer is designed to operate with 220 volts on the primary, and it will produce 24 volts at the secondary terminals. The volt-ampere rating of the transformer is 48 VA. This means that a maximum of 2 amperes can be drawn from the transformer's secondary (2 amperes $\times$ 24 volts = 48 VA).

SUMMARY

- The turns ratio of a transformer determines the relationship between input and output voltage. Np/Ns = Ep/Es.
- Power losses in a transformer are due to three main causes: hysteresis, eddy current, and IR losses.
- Transformers are rated in VA (volt-amperes) rather than power.

PRACTICAL EXPERIENCE

Required equipment   Three-phase distribution panel (208 V–240 V), voltmeter.

Procedure
1. Measure and record the individual voltages phase A to B, phase B to C, and phase C to A.

   A to B _____ volts

   B to C _____ volts

   C to A _____ volts

2. Measure the voltage phase A to neutral. _____ volts

3. Measure the voltage phase B to neutral. _____ volts

4. Measure the voltage phase C to neutral. _____ volts

5. Is the power panel being fed from a delta- or wye-connected secondary?

REVIEW QUESTIONS

1. The voltage ratio of a transformer is _____ proportional to the turns ratio.

2. The current ratio of a transformer is _____ proportional to the turns ratio.

3. The power supplied to the primary is approximately _____ to the power taken from the secondary.

4. The amount of power loss in transmission lines will be (higher, lower) _____ if the power is transmitted at higher voltages.

5. Why are low-voltage transformers used in air-conditioning systems?

6. A transformer has 800 turns on its primary and 400 turns on its secondary. The primary voltage is 200 volts. What is the secondary voltage?

7. In the transformer in question 6, if 4 amperes is drawn in the secondary circuit, how much current flows in the primary?

8. What are the three main losses in a transformer?

9. Which transformer loss is related to the motion of molecules?

10. What ratings are usually provided on control transformers?

Unit 13

Phase Shift and Power Factor

OBJECTIVES

Upon completion and study of this unit, you will understand that
- In ac circuits, current and voltage do not necessarily rise and fall together.
- Whenever a circuit is acting inductively or capacitively, there will be a phase shift between the current and voltage.
- Whenever there is a phase shift between current and voltage, power is no longer equal to $I \times E$. A third factor—the *power factor (pf)*—must be considered, hence,

$$W = I \times E \times \text{pf}$$

In Unit 10 it was shown that current and voltage in certain electric circuits do not rise and fall together. In an inductive circuit, such as a coil of wire, current lags behind (follows) voltage (Figure 13-1). Voltage passes through zero and rises in a positive direction, while current is still negative and follows behind the voltage. The current in such a circuit is called a *lagging current*.

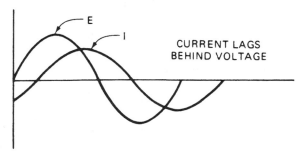

FIGURE 13-1 Voltage and current in an inductive circuit.

137

To better understand power in inductive circuits, it is well to take another look at power in resistive circuits. The current and voltage are in phase with each other. The power developed is equal to $I \times E$. In the graph in Figure 13-2, during the first time period, both the voltage and the current are positive. During the second time period, the voltage and the current are both negative. In algebra, a positive times a positive equals a positive ($+ \times + = +$), and a negative times a negative equals a positive ($- \times - = +$). During both time periods, therefore, the power is positive. This power is developed in the resistor as heat.

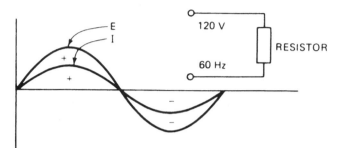

FIGURE 13-2   Voltage and current in a resistive circuit.

The power developed in a resistive circuit is shown in Figure 13-3. In a pure inductance circuit (one that does not contain resistance), current follows the voltage by 90°. Such a circuit is not possible. If it were, no power would be dissipated, and no heat would be developed.

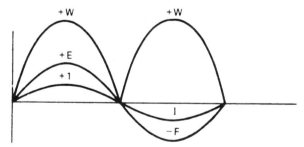

FIGURE 13-3   Power developed in a resistive circuit.

To help explain phase shift, a perfect inductor is considered. It must be noted, however, that this is for explanation only, since perfect inductors do not exist. All wires have resistance; all core materials have hystereses and eddy current losses.

In a perfect inductor, the current lags the voltage by exactly 90°. Figure 13-4 shows an iron core coil and provides a graph indicating that the current lags the voltage by 90°. The electrical energy provided by the 120-volt, 60-hertz source is converted into magnetic energy in the coil. The magnetic energy is converted back into electrical energy and is fed back to the incoming line. No power or energy is used in the coil; no heat is developed.

The power that is fed to the coil from the source is positive power. The power that is fed from the coil back to the source is negative power. In a pure inductance

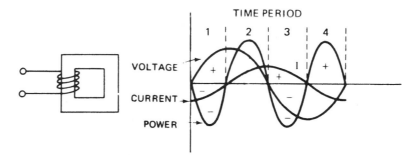

FIGURE 13-4   Current and voltage in a perfect inductor.

FIGURE 13-5   Positive and negative power in a pure inductance.

coil, they would be equal. An example of this is given in Figure 13-5, where current, voltage, and power are shown on the same graph.

1. In time period 1, a positive voltage times a negative current provides negative power ($+E \times -I = -W$).

2. In time period 2, a positive voltage times a positive current provides positive power ($+E \times +I = +W$).

3. In time period 3, a negative voltage times a positive current provides negative power ($-E \times +I = -W$).

4. In time period 4, a negative voltage times a negative current provides positive power ($-E \times -I = +W$).

The power fed to the circuit in time periods 2 and 4 is equal to the power fed from the circuit back to the line in time periods 1 and 3. No power is consumed in the inductance coil. This situation would only appear in a pure inductive circuit.

In actual practice, inductive circuits in air-conditioning and refrigeration applications are not pure inductive circuits. Some of the electric power that is fed to the circuit remains either as heat or is converted into mechanical power. A good example of a device converting electric power to mechanical power is a motor. In a circuit containing an electric motor, the current could lag the voltage by about 30°. This means that *some* of the power that is fed to the motor is not used but is fed back to the line. Actually, only 86.6% of the power fed to the circuit is used. This figure, 86.6%, is the *power factor* (pf) of the circuit. It is the cosine of the angle between the current and the voltage. A list of cosine values is presented in Table 13-1.

TABLE 13-1

Angle	Cosine	Power factor (%)
0	1.000	100.0
5	0.996	99.6
10	0.985	98.5
15	0.966	96.6
20	0.940	94.0
25	0.906	90.6
30	0.866	86.6
35	0.819	81.9
40	0.766	76.6
45	0.707	70.7
50	0.643	64.3
55	0.574	57.4
60	0.500	50.0
65	0.423	42.3
70	0.342	34.2
75	0.259	25.9
80	0.174	17.4
85	0.087	8.7
90	0.000	0.0

Graphically, this relationship is shown in Figure 13-6. Since current and voltage are not necessarily in phase with each other in circuits, this must be considered in the calculations of power. Power factor must also be considered in the calculations of power. The formula is

$$W = I \times E \times \text{pf}$$

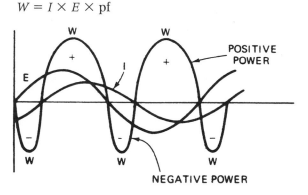

FIGURE 13-6   Power in a circuit containing resistance and inductance.

Since power factor is the cosine of the angle between the current and the voltage, the greater the angle is the smaller the power factor; the smaller the angle is the larger the power factor. When current and voltage are in phase, the angle is 0°; since the cosine of 0° is 1.0, the power formula is

$$W = I \times E \times \text{pf}$$

$$W = I \times E \times 1$$

$$W = I \times E$$

If all the power fed to the circuit were to be used in the circuit, no power would be returned to the source. This situation was shown in Figure 13-3.

When current and voltage are 90° out of phase with each other, as in a *pure* inductive circuit, the power factor (pf) is 0 (the cosine 90° is 0). The power formula becomes

$$W = I \times E \times \text{pf}$$

$$W = I \times E \times 0$$

$$W = 0$$

The power used in the circuit would be 0 since any number multiplied by 0 is 0. An example of this was shown in Figure 13-5 where the same amount of power supplied by the source as positive power is returned to the source as negative power.

In actual practice, of course, the power factor is seldom, if ever, as low as 0 or as high as 1.0. It is a percentage, such as given in Table 13-1. If the current lags the voltage by 30°, the power factor for 30° is 86.6% or 0.866.

The air-conditioning and refrigeration technician will seldom have to calculate or work with the power factor. It is necessary that the technician understand that the power formula does contain the pf component, in order that he or she not be confused when measuring voltage, current, and power.

CAPACITANCE CIRCUITS

The current in a capacitive circuit has characteristics exactly opposite to the current characteristics of inductive circuits. Current leads the voltage. In a perfect capacitor, current would lead the voltage (across the capacitor) by 90°. There is no such thing as a perfect capacitor, but nearly perfect capacitors in which current and voltage relations approach 90° are common. Figure 13-7 shows the current, voltage, and power relationships in a capacitive circuit.

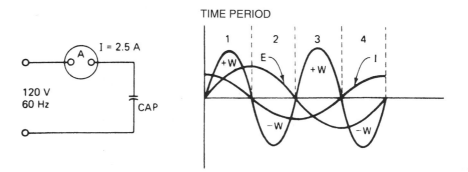

FIGURE 13-7 Current, voltage, and power relationship in a pure capacitive circuit.

1. In time period 1, current and voltage are positive; power is positive $(+I \times +E = +W)$.

2. In time period 2, current is negative, but voltage is positive; power is negative $(-I \times +E = -W)$.

3. In time period 3, current and voltage are negative; power is positive $(-I \times -E = +W)$.

4. In time period 4, current is positive, but voltage is negative; power is negative $(+I \times -E = -W)$.

The power fed to the capacitor in time periods 1 and 3 is equal to the power fed back to the line in time periods 2 and 4. No power is consumed in the capacitor.

Observe the relationship between Figures 13-5 and 13-7. In time period 1, the coil (Figure 13-5), is sending power back to the line (negative power). During this time period, the capacitor (Figure 13-7) is taking power from the line. During time period 2, the capacitor is sending power back to the line (negative power), while the coil is receiving power.

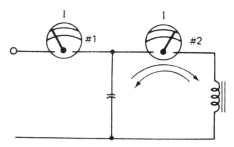

FIGURE 13-8 Low-line current with high-circulating current between capacitor and coil.

An interesting combination may be made with coils and capacitors. Assume that in their individual circuits, both the coil and the capacitor draw 2 amperes. If the two circuits are connected together in parallel, the line (source) has to supply very little power (Figure 13-8). Current circulates back and forth between the capacitor and the coil, as indicted on ammeter 2; little current is flowing in the line, as indicated on ammeter 1.

The power necessary to overcome the losses in the circuit is the only power supplied by the line. If perfect inductance coils and perfect capacitors were possible, no power at all would have to be supplied by the line. Current would be indicated in ammeter 2, whereas no current would be indicated in ammeter 1.

Capacitors are used by large manufacturing plants to reduce the incoming line current. Most manufacturing plants use a large number of motors in their operations. The motors are inductive and draw a lagging current. With a lagging current, some of the power that the power company is supplying is fed back to them. Because of losses in the transmission lines, the power company is losing power, and therefore money, because of the lagging current drawn by the manufacturing plant. To overcome this loss, the power company charges more for power supplied at lower power factors. This is not true for electrical power supplied to homes; the power company usually corrects for power factor at distribution points.

   Because of the higher charges presented to large manufacturing plants, it is
to the plant owner's advantage to correct the power factor and thus to reduce the
cost of operation. As an example, consider an electric motor drawing 900 volt-
amperes from a 208-volt power source (Figure 13-9). The uncorrected line current would be 4.33 amperes.

$$I = \frac{VA}{E}$$

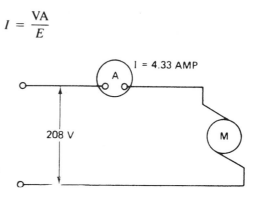

FIGURE 13-9   Motor drawing current with a lagging power factor.

$$I = \frac{VA}{E}$$

$$I = \frac{900}{208}$$

$$I = 4.33$$

By measurement, it is found that the power factor is equal to 0.866, indicating that
the current is lagging the voltage by 30°. The power factor may be determined by
measuring the current with an ammeter, the voltage with a voltmeter, and the
power with a wattmeter; the wattmeter reading is then divided by $I \times E$ or VA.

$$\text{pf} = \frac{W}{VA}$$

The circuit connections for determining the power factor are shown in
Figure 13-10.

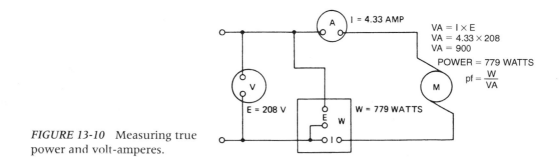

FIGURE 13-10   Measuring true
power and volt-amperes.

$$pf = \frac{W}{VA}$$

$$pf = \frac{779}{208 \times 4.33}$$

$$pf = \frac{779}{900}$$

$$pf = 0.866$$

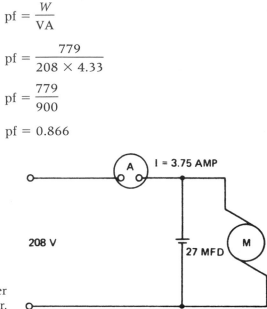

FIGURE 13-11 Correction of power factor by the addition of a capacitor.

The power factor of this circuit could be improved and the line current reduced by adding a capacitor in parallel with the motor (Figure 13-11). With only the addition of the capacitor, the line current could be reduced from 4.33 to 3.75 amperes; there would be no difference in the operation of the motor. The power company would be supplying 780 volt-amperes, and the motor would be using 780 watts.

This example covers the calculations of a single, low-horsepower motor. The example is typical for all the motors in a complete manufacturing plant. The explanation is given in order that the air-conditioning technician might have a better understanding of power factor and power-factor correction. It is not the job of an air-conditioning technician to correct power factor.

The capacitors that are used with compressor motors contribute to a better power factor although they are not included for this purpose. Capacitors are used in the start winding circuits to increase starting torque; the power-factor improvement that is obtained is of secondary importance.

VOLTAGE AND CURRENT MEASUREMENTS

Whenever the voltage and current are out of phase with each other in an ac circuit, their indications do not seem to follow the laws of series or parallel circuits. The laws of series and parallel circuits are true, but electrical measuring instruments provide a reading of effective values, rather than the instantaneous values on which the laws are based.

Consider the circuit of Figure 13-12: a resistor of 43 ohms and a capacitor of 100 microfarads are connected in series. The voltmeter across the capacitor indi-

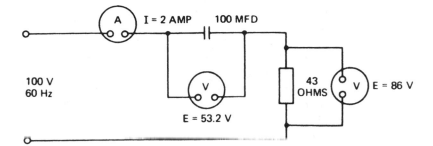

FIGURE 13-12   RC circuit $E_C + E_R$ is greater than the supply voltage.

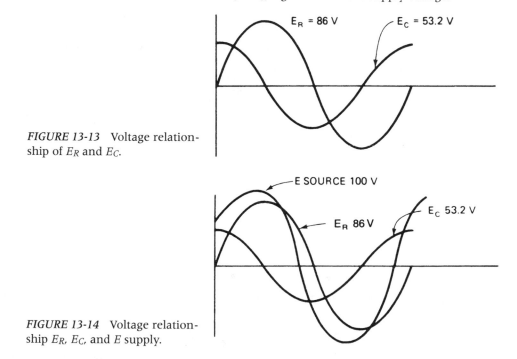

FIGURE 13-13   Voltage relationship of E_R and E_C.

FIGURE 13-14   Voltage relationship E_R, E_C, and E supply.

cates 53.2 volts, and the voltmeter across the resistor indicates 86 volts. It seems that the sum of these two voltages is not the same as the supply voltage of 100 volts (53.2 + 86 = 139.2). The waveforms of the two voltages are shown in Figure 13-13. Note that when the voltage across the resistor E_R is positive, the voltage across the capacitor E_C is first positive and then negative (one alternation of E_R). If these two waveforms are added together, the result is the supply voltage of 100 volts (Figure 13-14).

In a series circuit, the same current flows through each component. This law is true even if the current and voltage are not in phase with each other.

In parallel circuits, the same voltage appears across each component. This law is true for circuits where the current and voltage are not in phase with each other.

In parallel circuits, the line current equals the sum of the currents in the branches. This law is also true in ac circuits, but instantaneous values must be used.

Consider the circuit of Figure 13-15. A resistor, a capacitor, and a coil are connected in parallel. Ammeters connected in series with each component indicate the current flow through the component. The ammeter in series with the line indicates the total line current. It is not possible simply to add up the currents in the branch circuits to get the line current. In Figure 13-16, the waveforms of the individual currents are shown. The total line current is the sum of the three waveforms of current. Since the current in the capacitor and the current in the coil have polarities exactly opposite each other, the capacitive current can be added to the coil current arithmetically. This leaves a current of 10 amperes acting inductively: $+(20\ I_C) - (30\ I_L) = -10\ I_L$.

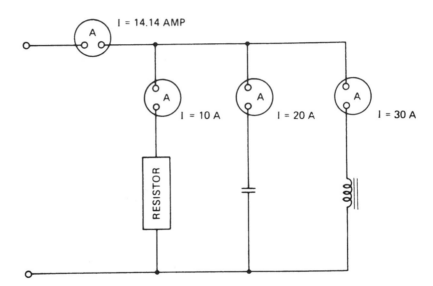

FIGURE 13-15   Parallel circuit containing resistance, capacitance, and inductance.

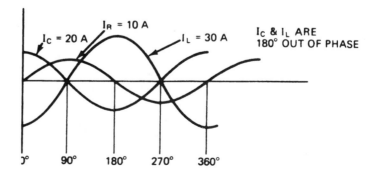

FIGURE 13-16   Phase relation of current in the resistor, capacitor, and coil.

The current waveforms are now as shown in Figure 13-17. When the two currents are added together, the final result is as shown in Figure 13-18.

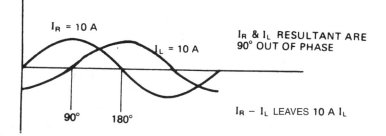

FIGURE 13-17 Resistive current and resultant inductive current.

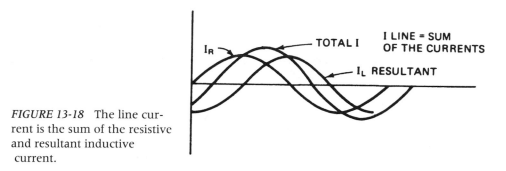

FIGURE 13-18 The line current is the sum of the resistive and resultant inductive current.

MOTOR CIRCUITS

The compressor motor start-winding circuit of most large air-conditioning systems includes a capacitor in series with the start winding (Figure 13-19). The voltage across the capacitor might be as high as 320 volts. The capacitor voltage and start-winding voltage do not add up to the supply voltage of 220 volts.

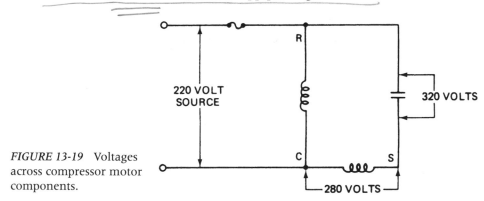

FIGURE 13-19 Voltages across compressor motor components.

An air-conditioning and refrigeration technician will often make voltage and current measurements in circuits where there is a phase difference between the current and the voltage. It is important that the meter indicators do not cause any confusion.

The service technician will not have to deal with current or voltage waveforms. This information has been presented so that the technician will have a better insight into circuit operations for troubleshooting procedures.

SUMMARY

- In ac circuits, current and voltage seldom are in phase with each other.
- Power in ac circuits is equal to $I \times E \times$ pf. *= True Power*
- The power factor (pf) is equal to the cosine of the angle between the current and voltage.

PRACTICAL EXPERIENCE

Required equipment   Air conditioner with a capacitor-run compressor motor, voltmeter. Access to the compressor motor components is required.

Procedure

1. Connect the A/C unit to the voltage specified.

2. Measure and record the input voltage._____volts

3. Measure and record the voltage across the run winding._____volts

4. Measure and record the voltage across the start, or AUX, winding._____volts

5. Measure and record the voltage across the run capacitor.

6. Do the voltage readings of steps 5 and 6 indicate a phase relation between the two voltages of a series circuit?

REVIEW QUESTIONS

1. In a circuit containing resistance and inductance, the line current will lag behind the voltage.   T_____   F_____

2. Whenever the current and voltage are out of phase with each other, the total power developed in a circuit is equal to $I_T \times E_T$.   T_____   F_____

3. The cosine of a 45° angle is 0.866.   T_____   F_____

4. The current is always in phase with the voltage across a resistance.   T_____   F_____

5. The current flow in a capacitor leads the voltage across the capacitor by 90°.   T_____   F_____

6. Pure inductance circuits are commonly found in air-conditioning circuits.   T_____   F_____

7. The formula for power in an ac circuit is $W = I \times E \times 2$.   T_____   F_____

8. To find the total applied voltage in an ac series circuit containing inductance and capacitance, the voltages across the components are simply added together.   T_____   F_____

9. The power factor in an ac circuit is often over 200%.   T_____   F_____

10. The power factor of a circuit containing only pure capacitance would be 0.50.   T_____   F_____

Unit 14

Electric Motors

OBJECTIVES

Upon completion and study of this unit, you will be familiar with the principles of and the differences and similarities between
- Induction motors.
- Shaded pole motors.
- Split-phase motors.
- Capacitor motors.
- Polyphase motors.
- Two-speed, single-phase systems.
- Two-speed, three-phase systems.
- Variable-speed, three-phase systems.

Most air-conditioning and refrigeration systems operate through the use of a compressor of one type or another, driven by an electric motor. The most common type of motor used to drive these compressors is the induction motor.

An *induction motor* operates on a basic principle of electricity: an induced current opposes the field that produced it. This was covered in Unit 12, where the secondary current of a transformer opposed the field of the primary. The back EMF of the primary tends to be reduced, allowing for more current in the primary. In this way, power in the secondary was provided for by increased power in the primary. Refer to Figure 12-4.

An induction electric motor is actually nothing more than a special type of transformer. The primary of the motor transformer is wound on the stationary parts, called the *stator*. The secondary is made up of copper bars shorted together on the rotating part of the motor, called the *rotor*.

The primary windings on the stator look very much like most transformer windings. The secondary winding on the rotor consists of copper bars on an iron core. The copper bars are all connected together. This essentially provides a combination of short-circuited single-turn windings. Figure 14-1 shows a typical rotor in which the core is made up of die-cut laminations. These laminations reduce eddy currents in the iron core of the rotor.

Whenever an ac voltage is connected to the primary of a transformer, a voltage is induced in the secondary. The secondary induced voltage is in direct opposition to the primary supply voltage. This opposition usually results in a rotation of the motor shaft.

FIGURE 14-1   Rotor. (Courtesy of BET Inc.)

Without some form of electric direction, the rotor may turn in either direction, clockwise (cw) or counterclockwise (ccw) (Figure 14-2). If the rotor were not started in some direction, the motor would sit like a transformer with a shorted secondary—it would hum but not move. In part of alternation time period 1, current flows through the stator to provide for north and south poles. The induced voltage and the current flow in the rotor produce a magnetic field in the secondary winding in opposition to the magnetic field in the primary winding. The system is in balance; the rotor does not rotate. Some direction action is needed.

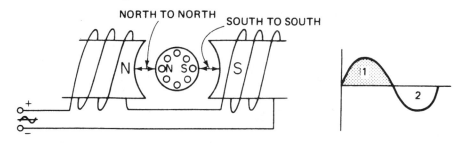

FIGURE 14-2   Magnetic field set up in rotor due to stator magnetic field.

In part of the second alternation of the applied voltage time period 2, the current flow and the polarity of the magnetic field of the primary change. The current flow and magnetic polarity in the secondary also change (Figure 14-3). Although the magnetic field of the stator and rotor change, the rotor still does not turn since the opposition of magnetic poles is direct, trying to force the rotor toward the shaft. center.

When the motor shaft is started in either direction, a time-delay imbalance is created, allowing the motor to continue to turn in the direction in which it was started. The motor continues to turn after it is started because of the repulsion between the stator poles and the poles set up in the rotor (Figure 14-4). As long as electric power is applied to the stator winding of the motor, it continues to rotate in the direction in which it was started. A continuous shift in the magnetic field about the rotor takes place. Current flow in the rotor results from two actions: transformer action of the primary, as previously discussed, and generator action as the copper bars in the rotor cut through the primary magnetic field.

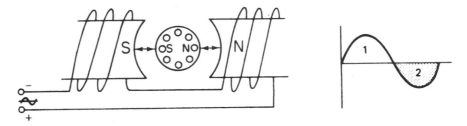

FIGURE 14-3   In negative alternation, the magnetic field of the stator and rotor both change direction.

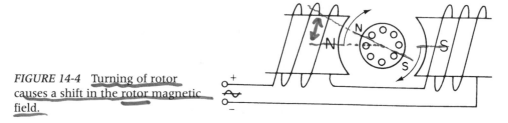

FIGURE 14-4   Turning of rotor causes a shift in the rotor magnetic field.

SHADED-POLE MOTORS

One method used to get small motors ($\frac{1}{4}$ horsepower or less) started is to provide shaded poles in the motor (Figure 14-5). There is a slotted area in the two pole pieces with shorted turns of copper wire in the slots. The transformer action involving these shorted turns starts the motor in the desired direction.

The magnetic polarity relationships of the rotor and stator of a shaded-pole motor are shown in Figure 14-6. This relationship exists just after the current flow in the motor coil passes through zero.

The current flow in the shaded pole (shorted turn) sets up a magnetic field in opposition to the magnetic field that produced it. The magnetic field set up by the shaded pole is of the same polarity as the magnetic field set up by the rotor currents. The magnetic field of the shorted turn is not as strong as the magnetic field of the main coil, but it is sufficient to cause a delay in the magnetic field set up in the shaded area.

The end result is a rotating magnetic field about the stator of the motor. The relationship is shown in Figure 14-7 and is as follows:

1. The current in the main winding passes through zero and starts to increase. A north pole is started in the main pole area. The opposing effect in the shaded area produces a south pole at the shaded-pole area.

2. As the main field builds up toward a maximum north pole, the shaded area becomes a weak north pole. The effect of the shaded pole is decreasing.

FIGURE 14-5 Shaded-pole motor. (Courtesy of BET Inc.)

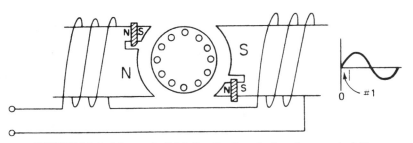

FIGURE 14-6 Magnetic field distribution during time period #1.

3. The main pole is at maximum. The effect of the shaded pole is zero.

4. The strength of the main pole decreases from a maximum north pole. The shaded area becomes a north pole.

5. The main pole decreases in strength to zero. The shaded area stays a north pole.

6. The main pole changes polarity to a south pole. The shaded area remains a north pole.

As long as the magnetic field is rotating around the stator, the resulting current flow in the rotor causes the rotor to follow. The direction of rotor rotation is always in the direction of the shaded pole (Figure 14-8).

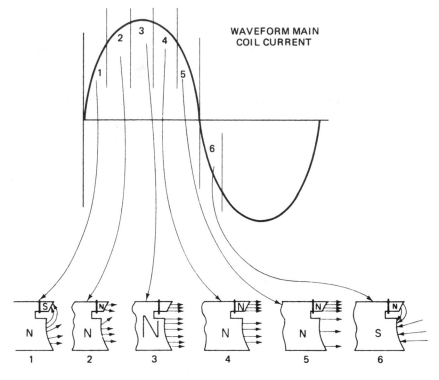

FIGURE 14-7   Stator field during time periods 1 through 6.

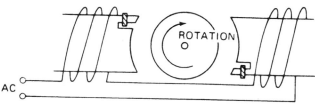

FIGURE 14-8   Rotation is
towards shaded pole.

Motor Direction (Shaded Pole)

Many shaded-pole motors have a shaft extending out of end bells on both sides of
the rotor. With a double-shaft motor, either clockwise or counterclockwise rotation is obtained, depending on which shaft is used to drive the load (Figure 14-9).
In many cases, it is possible to obtain reversed rotation of a shaded-pole motor with
a single shaft (extending from only one end). If the end bells of the motor are
interchangeable, the motor may be reversed simply by reversing the rotor. The end
bells are removed, and the motor is reassembled with the shaft extending from the
opposite end of the motor (Figure 14-10).

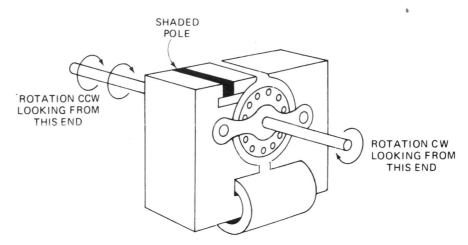

DUAL SHAFT SHADED POLE MOTOR

SHADED
POLE

ROTATION CCW
LOOKING FROM
THIS END

ROTATION CW
LOOKING FROM
THIS END

FIGURE 14-9 Dual shaft motor—shafts rotate in opposite direction when looking at shaft end.

except capacitor-start type.

SPLIT-PHASE MOTORS

Split-phase motors are usually used for applications requiring from $\frac{1}{6}$ to $\frac{3}{4}$ horsepower. The starting torque of split-phase motors is low to moderate compared with other motor types, except shaded-pole motors. Split-phase motors are usually found in service for driving fans, centrifugal pumps, conveyors, and refrigerator compressors. The split-phase motor is the most common type used in home refrigerator compressor units.

The rotating magnetic field needed to start split-phase motors is obtained through the use of two stator windings in the motor, a main winding, and a start winding. Each of these windings has different inductance and resistance values. Current flows in the two windings are not in phase with each other (Figure 14-11). The current in the main winding (coil A) lags the current in the start winding (coil B). A rotating magnetic field is set up in the stator, and the motor rotates in the direction in which the field rotates.

After a split-phase motor has started its rotation and as its speed approaches normal operating speed, the start winding is no longer needed. A mechanical centrifugal switch is often used to disconnect electrically the start winding from the power source, usually at about two-thirds of top speed (Figure 14-12). The motor continues to rotate after the start winding has been electrically disconnected.

Motor Direction (Split-Phase)

The direction of rotation of most split-phase motors may be reversed by interchanging the connections of the windings. The motor windings are marked S_1 and S_2 for the start winding and R_1 and R_2 for the run winding. In Figure 14-12a, the winding connection is S_1 to R_1 and S_2 to R_2; the motor rotates clockwise. In

TO REVERSE DIRECTION
DISASSEMBLE AND
REVERSE SHAFT

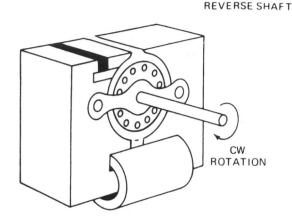

CW
ROTATION

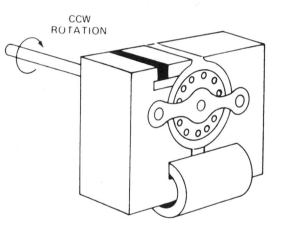

CCW
ROTATION

FIGURE 14-10   With a single shaft motor, end bells must be interchanged and rotor installed at opposite side, to reverse direction.

Figure 14-12b, the connection is S_1, to R_2 and S_2 to R_1; the motor rotation is counterclockwise.

The centrifugal switch is in series with the (S) winding of the motor (Figure 14-13). Although bearings often wear out in motors using centrifugal switches, the switch and switch mechanism have proved to be a common source of trouble. The contacts of the switch may become burned and pitted, preventing good electrical contact. Dirt and grit may interfere with the mechanical operation of the switch mechanism, causing the switch to remain open after the motor has been stopped.

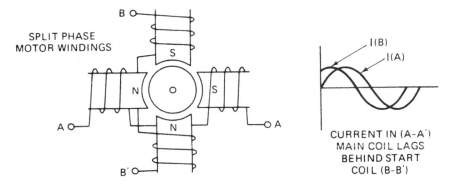

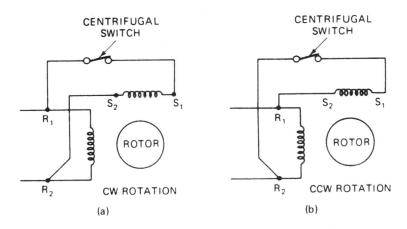

FIGURE 14-11 Split-phase motor winding produces rotating magnetic field.

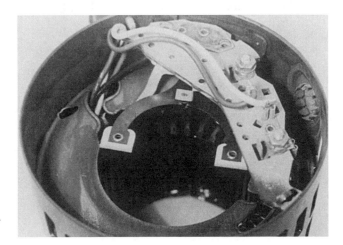

FIGURE 14-12 Changing direction of rotation of a split-phase motor.

FIGURE 14-13 Motor
stator–centrifugal switch.
(Courtesy of BET Inc.)

- motor could also be stalled due to jambing. In either case turn off!

If the switch contacts are open and an attempt is made to start the motor, the shaft will not rotate. The motor will hum but will not turn. If this should happen, it is necessary to disassemble the motor and to clean the mechanical portion of the centrifugal switch. The contact surfaces of the electrical switch should be inspected for excessive pitting. Minor pitting may be removed and the contact surface returned to operating condition with fine sandpaper. Care must be taken to remove all sand and sanded contact material from the motor. If in doubt, the centrifugal switch should be replaced as an assembly.

CAPACITOR MOTORS (SPLIT-PHASE) → *Single Phase type motors*

Capacitor-type, split-phase motors are used in applications where higher starting torque and/or higher horsepower ratings are required. Common capacitor-type, split-phase motors are available from $\frac{1}{3}$ to 10 horsepower. There are three different types of capacitor split-phase motors common to the air-conditioning and refrigeration industry.

1. Permanent split-capacitor (PSC) motor
2. Capacitor-start motor
3. Capacitor-start, capacitor-run motor

Permanent Split-Capacitor Motors

The permanent split-capacitor motor is usually of the fractional horsepower size, $\frac{1}{20}$ to $\frac{1}{2}$ horsepower. These motors are often used in place of shaded-pole motors because they are somewhat more efficient. The capacitor remains in series with the auxiliary winding of the motor at all times. An electrical circuit of this motor is shown in Figure 14-14.

Good power factor!   *Causes phase shift between windings.*

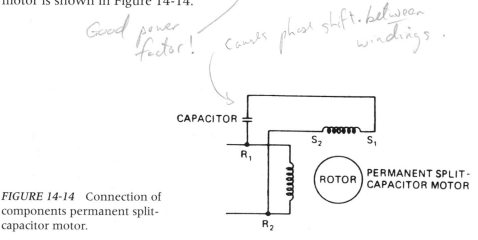

FIGURE 14-14 Connection of components permanent split-capacitor motor.

Capacitor-Start Motors

Capacitor-start motors are readily available from $\frac{1}{4}$ to 10 horsepower. They are most often used in applications in which high starting torque is required.

Capacitor-start motor operation is very similar to that of the split-phase induction motor with one exception. The capacitor-start motor has an electrolytic capacitor in series with the start winding, sometimes called the auxiliary winding. A centrifugal switch or some other control device is used to remove electrically the capacitor and the start winding from the circuit after the motor reaches approximately $\frac{3}{4}$ of operating speed. The diagram of a capacitor-start motor, Figure 14-15, is very similar to that of the split-phase motor, Figure 14-12; the only difference is the addition of the capacitor.

high start torque.

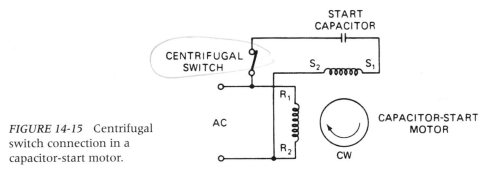

FIGURE 14-15 Centrifugal switch connection in a capacitor-start motor.

Motor Direction (Capacitor-Start) The direction of rotation of capacitor-start motors may be changed by reversing the electrical relationship of the two windings. If the motor connections in Figure 14-15 are taken as standard, then the motor direction is reversed by the motor connections shown in Figure 14-16. The method that the technician uses to reverse motor direction depends on what access is available to the winding connections.

centrifugal switch

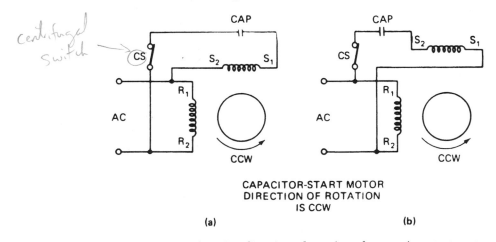

CAPACITOR-START MOTOR
DIRECTION OF ROTATION
IS CCW

(a)   (b)

FIGURE 14-16 Changing direction of rotation of a capacitor-start motor.

A common trouble area in capacitor-start motors is the starting mechanism, which may be a centrifugal switch or some other control device such as a relay. More information on motor starting circuits is given in Units 15 and 16. A second trouble area is the start capacitor. Failure of the start capacitor, usually an electrolytic one, may be due to an open or a short.

Capacitor-Start, Capacitor-Run Motors

The capacitor-start, capacitor-run motor is essentially a combination of the permanent split-capacitor motor and the capacitor-start motor. It is a common motor in air-conditioning systems. The capacitor-start, capacitor-run motor provides both the high starting torque of the capacitor-start motor and the efficiency (as well as good power factor) of the permanent split-capacitor motor.

A diagram of a capacitor-start, capacitor-run motor is shown in Figure 14-17. High starting torque is obtained through the use of an electrolytic (start) capacitor, C_2, in parallel with the permanent oil-filled (run) capacitor, C_1. When power is applied to the motor and as shaft rotation approaches operating speed, the control switch, S_1, opens. The large-value capacitor, C_2, is removed from the circuit, and total current draw by the motor is reduced. The permanent capacitor, C_1, remains in the circuit, providing for constant torque as well as improving the overall power factor of the motor.

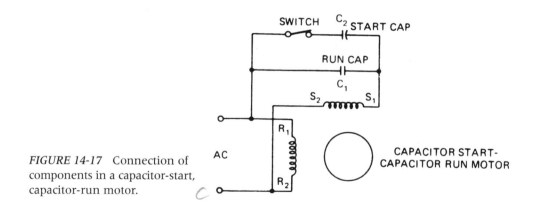

FIGURE 14-17   Connection of components in a capacitor-start, capacitor-run motor.

The capacitor-start (capacitor-run) motor is probably the most common type of motor used in the air-conditioning hermetic compressor system. The run and start windings of the motor are connected together internally, and three connections are brought out to motor terminals. These terminals are marked R for run winding, S for start winding, and C for common connection. This arrangement is shown in Figure 14-18.

There are times when a service technician may have to determine the proper terminal marking for a single-phase compressor. An ohmmeter is all that is necessary in making this determination. The following steps should be taken when checking for motor terminal connections:

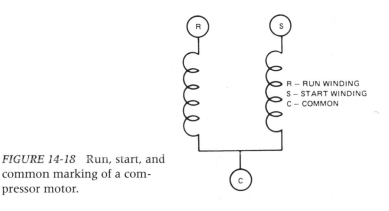

R – RUN WINDING
S – START WINDING
C – COMMON

FIGURE 14-18 Run, start, and common marking of a compressor motor.

1. Mark the terminals 1, 2, and 3.

2. Measure the resistance between each set of terminals.

3. The resistance between one set of terminals will be higher than the resistance between the other two sets. The terminal not included in the highest resistance measurement is the common terminal. Mark that terminal C.

4. Measure the resistance between the common terminal and the other terminals.

5. The indication of lower resistance from a terminal to common C determines the run (R) winding.

6. The indication of higher resistance from a terminal to common C determines the start (S) winding. └─ *more coils - more voltage across coil.*

An example of winding resistance is shown in Figure 14–19, which shows the ohmmeter measurements.

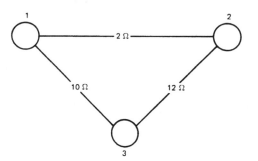

FIGURE 14-19 Measuring resistance in order to determine R, S, and C terminals.

1. The highest resistance measured, 12 ohms, is measured between terminals 2 and 3. Terminal 1 must be the common terminal and should be marked C.

2. The resistance between terminals 1 and 2 is 2 ohms, and the resistance between terminals 1 and 3 is 10 ohms.

3. The lowest resistance, 2 ohms, indicates that terminal 2 is the run (R) winding.

4. The next higher resistance, 10 ohms, indicates that terminal 3 is the start (S) winding.

REPULSION-START, INDUCTION-RUN MOTORS

Repulsion-start, induction-run motors are not common in the air-conditioning and refrigeration industry today. Older operating units may be found with this type of motor still in use.

The repulsion-start induction motor includes coil windings on the rotor (Figure 14–20). The individual ends of the coil winding are connected to a commutator. A centrifugal mechanism controls the start of the motor, where a set of brushes connects the ends of some of the rotor windings together. As the motor approaches operating speed, the centrifugal device completely shorts the commutator segments together through the use of a shorting ring. The rotor essentially becomes a squirrel-cage rotor. The high manufacturing cost of this motor, compared to other types of motors, ended its popularity.

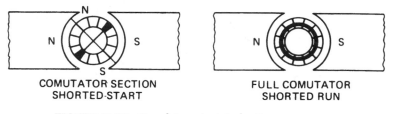

COMUTATOR SECTION   FULL COMUTATOR
SHORTED-START   SHORTED RUN

FIGURE 14-20   Repulsion-start, induction-run motor.

POLYPHASE MOTORS (THREE-PHASE)

Polyphase motors are simple in construction. They do not require starting windings, centrifugal switches, or start capacitors.

When a three-phase voltage is applied to the stator of a three-phase motor, a rotating magnetic field is automatically present around the stator (Figure 14–21). A set of magnetic pole pairs is indicated by the numbers 1 ①, 2 ②, and 3 ③. With the application of the three phases of voltage to the windings of the poles, a rotating field results around the face of the motor stator.

First, pole 1 is a north pole, then pole 2, and then pole 3. Pole ① then becomes a south pole as the current changes direction. Poles ①, ②, and ③ were south poles in order as 1, 2, and 3 were north poles. The magnetic field continuously rotates around the stator face. A squirrel-cage rotor rotates in a magnetic field of this type. No other starting system is needed.

Winding Connection

The windings of three-phase motors are connected in either the wye or delta configuration, depending on voltage and current requirements. As a current-limiting scheme, three-phase motors are sometimes started in the wye configuration and are switched into the delta configuration after the motor comes up to speed. In this manner, the starting current is limited. Figure 14-22 shows examples of wye and delta connections.

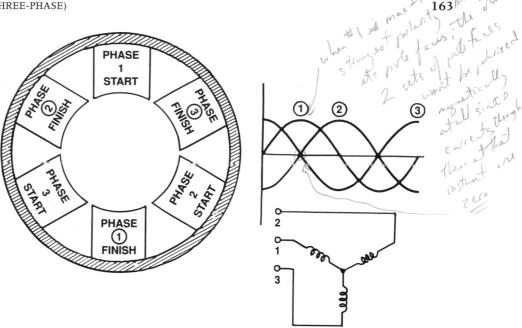

FIGURE 14-21   Three-phase motor.

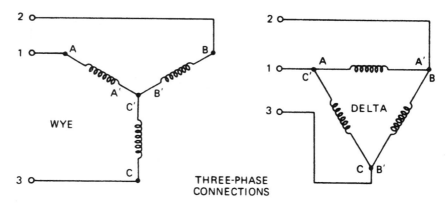

FIGURE 14-22   Wye (Y) and delta (Δ) connections.

Direction Control of Three-Phase Motors

To reverse the direction of a three-phase motor, it is only necessary to interchange any two of the three supply voltage lines. In this manner, the direction of rotation of the magnetic field is reversed and, consequently, so is the direction of rotor rotation.

MOTOR SPEED

The speed of an ac motor is controlled by the number of poles of the motor and the frequency of the applied voltage. By formula, the speed is

$$\text{rpm} = \frac{2F \times 60}{N}$$

where rpm = motor speed in revolutions per minute

2 = factor (it takes two poles to complete a magnetic circuit)

F = frequency of power in hertz (cycles per second)

60 = conversion from cycles per second to cycles per minute

N = number of poles in the motor

The speed of a two-pole motor is

$$\text{rpm} = \frac{2F \times 60}{N}$$

$$\text{rpm} = \frac{2 \times 60 \times 60}{2} = \frac{7200}{2}$$

$$\text{rpm} = 3600$$

This is the *synchronous* speed of the motor, the speed at which the motor would operate if there were no slippage.

The full-load speed of the motor is 3600 rpm minus approximately 150 rpm of slippage, for an *operating* speed of 3450 rpm.

Four-pole motor operation becomes

$$\text{rpm} = \frac{2F \times 60}{N}$$

$$\text{rpm} = \frac{2 \times 60 \times 60}{4}$$

$$\text{rpm} = 1800$$

Considering slippage, the motor speed becomes 1725 rpm.

SLIPPAGE

Slippage is the difference between the synchronous speed, 3600 rpm for a two-pole motor or 1800 rpm for a four-pole motor, and the actual operating speed. For example, in a four-pole motor the magnetic field rotates around the stator at a synchronous speed of 1800 rpm. To develop a magnetic field in the rotor of the motor, the short-circuited copper bars of the rotor must cut through the magnetic field of the stator. With the four-pole stator, the field rotates at 1800 rpm. If the rotor turned at 1800 rpm, the copper bars of the rotor would not cut through the magnetic field of the stator, but would be rotating at exactly the same speed. This situation cannot be! Since there would be no current in the rotor's copper bars, there would be no rotor magnetic field. There would be no power to cause the rotor to move. The rotor would slow down.

When the rotor turns at a speed less than the synchronous speed, the rotor cuts through the magnetic field of the stator. Current flows in the rotor bars, and a magnetic field is developed in the rotor. The reactions of the stator and rotor magnetic fields cause the rotor to turn.

When a motor is operated at no load, the slippage will be small. The rotor cuts through a limited amount of the stator magnetic field. When a load is put on the motor, the motor slows down (more slippage). The rotor cuts through a larger portion of the stator magnetic field. Higher current flows in the rotor, thus providing the power to turn the load.

Note: The motor is essentially a transformer. The stator is the primary and the rotor the secondary. When current increases in the secondary (rotor), the current will increase in the primary (stator). When more power is needed, more power is drawn.

Induction motors turn at a speed below synchronous speed; they have slippage. A type of motor that runs without slippage is called a *synchronous motor*. A synchronous motor has a different rotor than does an induction motor. Usually, a separate winding is placed on the rotor. This winding is fed dc current through slip rings to produce a magnetic field. (No slippage is needed.) Originally, the synchronous motor found little application in air conditioning. However, some of the new variable-speed, three-phase compressors are using permanent magnets in the rotor of the compressor motor. This type of motor would be considered a synchronous motor. The motor would turn at or close to synchronous speed.

TWO-SPEED COMPRESSOR MOTOR

The increase in energy costs has necessitated the development of increased efficiency air-conditioning systems. One development that has provided for increased efficiency is the two-speed compressor. When the demand is high, the compressor operates at high speed; when the demand is low, the compressor is switched to low speed.

Demand control may be obtained through the use of a standard two-stage thermostat similar to the one shown in Figure 18-20.

Speed control of the compressor motor is obtained by changing the winding connections from two-pole operation at high speed to four-pole operation at low speed. It has already been shown that at 60 hertz a two-pole motor operates at a speed of slightly less than 3600 rpm. A four-pole motor operates at slightly less than 1800 rpm.

Low-Speed Operation

Figure 14-23 shows the motor winding connections for a single-phase, four-pole, low-speed compressor. Consider an instant in time during a cycle of the input voltage. Line $L1$ is negative, and line $L2$ is positive. Using the left-hand rule (see Figure 2-16) and the direction of electron flow, the polarity of the motor coils may be determined. The direction of electron flow is indicated by the arrow in Figure 14-23. The polarity is shown in the diagram. The motor is four pole and will run at about 1725 rpm.

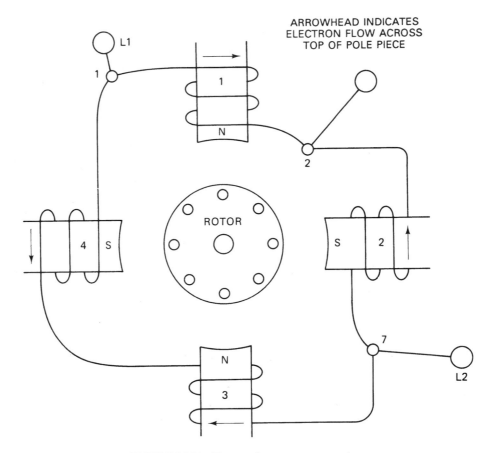

FIGURE 14-23   Four-pole motor connection.

High-Speed Single-Phase

When high-speed operation is called for, the connection of line $L2$ is changed from motor terminal 7 to motor terminal 2. At the same time, motor terminal 7 is connected to motor terminal 1. The connections are shown in Figure 14-24.

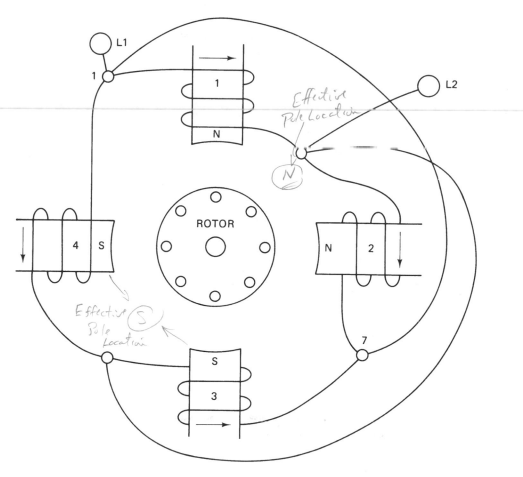

FIGURE 14-24 Four-pole motor connected for two-pole operation.

Again, consider the instant in time when $L1$ is negative and $L2$ is positive. The direction of electron flow is indicated by arrows in Figure 14-24. Poles 1 and 2 are north poles, and poles 3 and 4 are south poles. The motor effectively becomes a two-pole motor, with the two poles as shown in Figure 14-25. The motor speed is about 3450 rpm.

Start Windings

Separate start windings are used for low-speed, four-pole and high-speed, two-pole operation of the two-speed single-phase compressor. Selection of either the low-speed start winding or the high-speed start winding is made by the speed-control relays. The same start and run capacitors are used for either low- or high-speed operation.

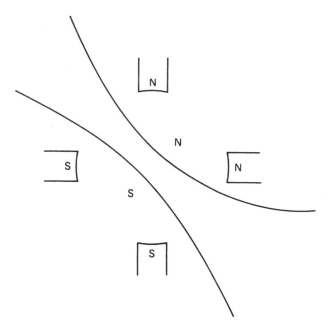

FIGURE 14-25   Four-pole combination for two-pole effect.

Single-Phase Wiring Diagrams

The control circuit connections for low-speed operation of the compressor are shown in Figure 14-26. The heavy lines indicate power connections through closed relay contacts. The low-speed contactor is energized during low-speed operation. The control circuit connections for high-speed operation are shown in Figure 14-27.

THREE-PHASE COMPRESSOR

The three-phase two-speed compressor operates in a manner similar to the single-phase two-speed compressor. The motor windings are switched from four-pole at low speed to two-pole at high speed. Again, a simple two-stage thermostat could be the control input.

Low-Speed Three-Phase

For low-speed operation, the stator windings of the compressor motor are connected in series as shown in Figure 14-28. Only half the windings are shown for the sake of simplicity. The second half of the motor would have the same winding connections as the first half. Note the different connection of the individual pole pieces 2 and 5. This is the end of the phase 3 winding, which would have started on the other side of the motor at pole pieces 8 and 11 if the complete circuit were shown.

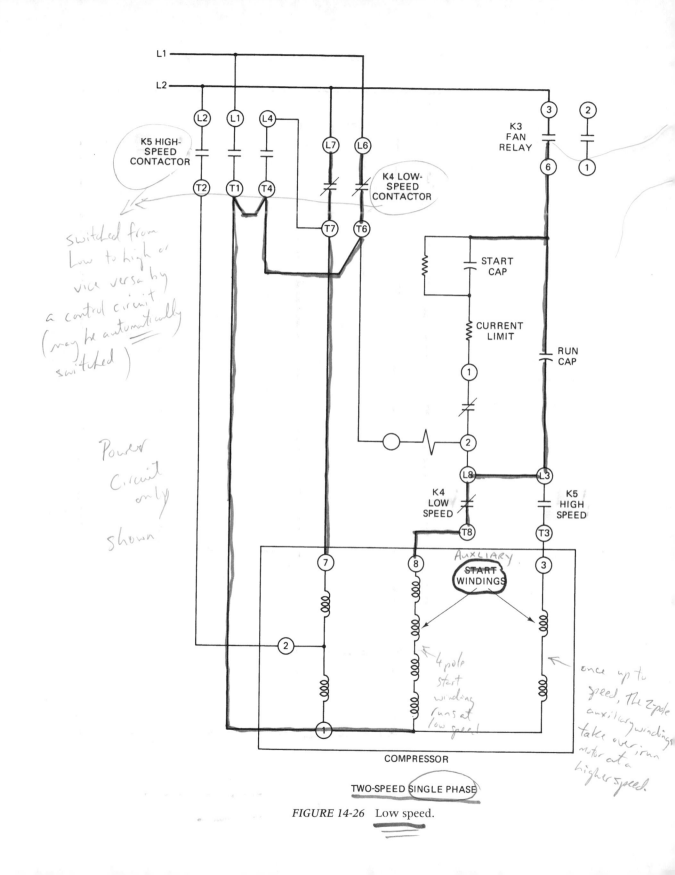

FIGURE 14-26 Low speed.

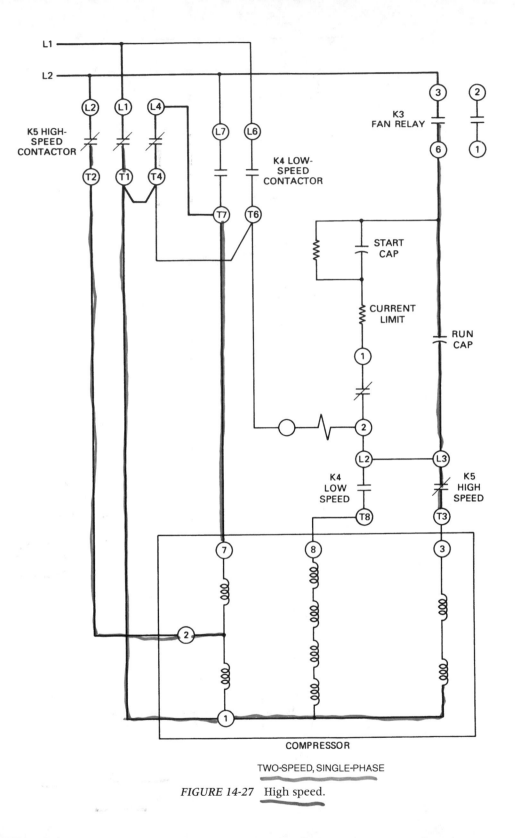

TWO-SPEED, SINGLE-PHASE

FIGURE 14-27 High speed.

The diagram Figure 14-28 shows the beginning of phase 1 and phase 2 connection to pole pieces 1 and 4 and 2 and 6, respectively. The ends of these phases are at the other sides of poles 4 and 10 and 9 and 12, respectively.

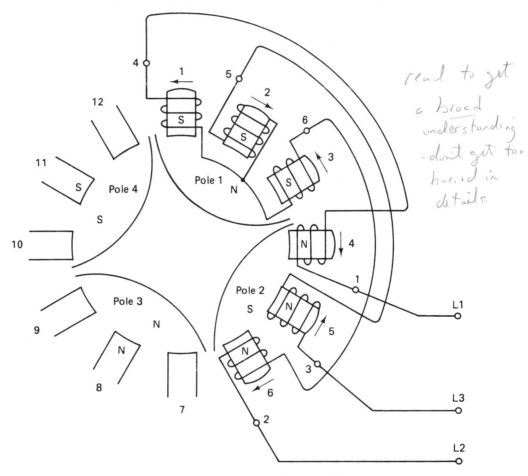

read to get a broad understanding - don't get too buried in details.

FIGURE 14-28   Three-phase low speed.

Consider a point in time, A, on the three-phase waveform (Figure 14-29). Here line L3 is at maximum negative. Line L1 is positive and moving in a negative direction. Line L2 is positive and moving in a positive direction. The small arrows in Figure 14-28 indicate the direction of electron flow across the top of the pole. The polarity of the pole is found using the left-hand rule, shown in Figure 2-16. Pole 1 is a maximum north pole and pole 2 is a maximum south pole. The connection of the windings form a four-pole motor.

In the pole pieces, pole 1 is a weak south pole and becoming weaker. Pole 2 is a maximum strength south pole, and pole 3 is a weak south pole and becoming stronger. Pole 4 is a weak north pole and becoming weaker. Pole 5 is a maximum north pole, and pole 6 is a weak north pole and becoming stronger. (See Figure 14-30).

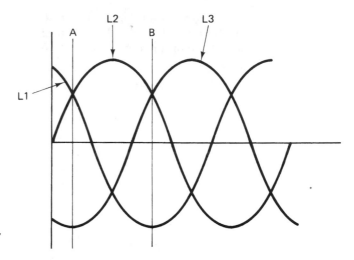

FIGURE 14-29 Three-phase sine waves: lines L_1, L_2, and L_3.

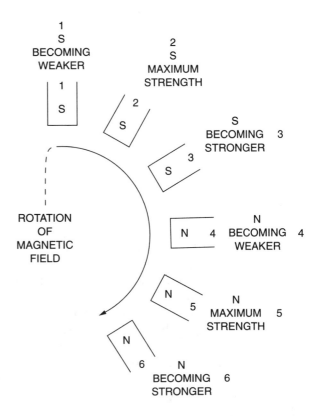

FIGURE 14-30 Magnetic field strength at time point "A" (Figure 14-29).

The direction of rotation of the magnetic field is clockwise, as will be the direction of motor rotation. This is a four-pole motor. The motor will run at about 1750 rpm. In Figure 14–31, a schematic diagram of the wiring is shown. In Figure 14-32, a wiring diagram is shown. Note the indication of closed contacts on the low compressor contactor, K4. The windings are in series. The direction of current flow is shown on the three sets of windings, Figure 14-31.

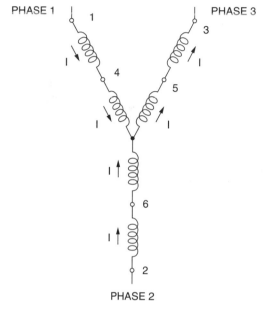

FIGURE 14-31 Schematic three-phase, low-speed with current direction (point A Figure 14-29).

The wiring diagram of a three-phase motor connected for high-speed operation is shown in Figure 14-33. The high compressor contactor, K5, contacts are shown closed here. The direction of current flow through the 1, 2, and 3 terminals is reversed with this connection.

Reversing the direction of current in poles 4, 5, and 6 changes them to south poles at time point A (Figure 14-29). One half of the motor becomes a north pole, while the other half becomes a south pole. This is effectively now a two-pole motor, as shown in the schematic diagram in Figure 14-34. Compare the direction of current flow in the windings. Figure 14–31 shows low-speed operation. Figure 14-35 shows high-speed operation.

When the low compressor relay K4 is de-energized and the high compressor relay K5 is energized, two actions take place. First, the direction of current flow through three of the coils is reversed. Second, the coils are connected in parallel rather than in series.

Compare the direction of current flow in the coils of Figure 14-31 versus Figure 14-35. The arrows indicating current direction change in terminals 1, 2 and 3. If the direction of current flow through the coil changes, the magnetic polarity of the coil changes.

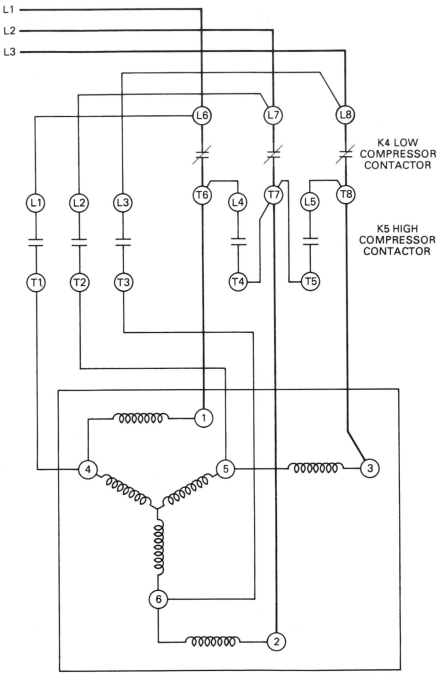

FIGURE 14-32 Three-phase, low-speed.

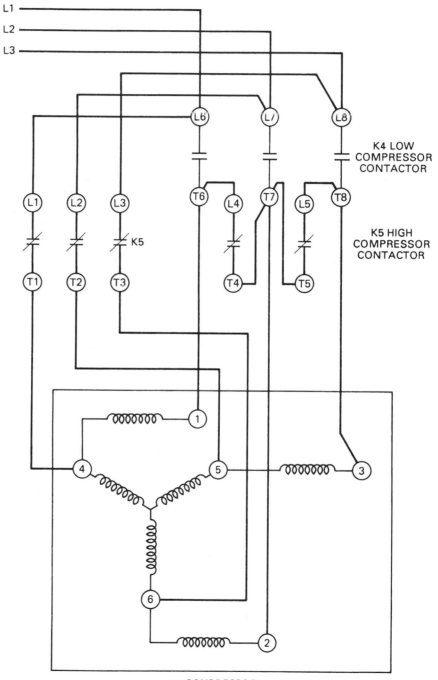

COMPRESSOR

FIGURE 14-33 Three-phase, high-speed.

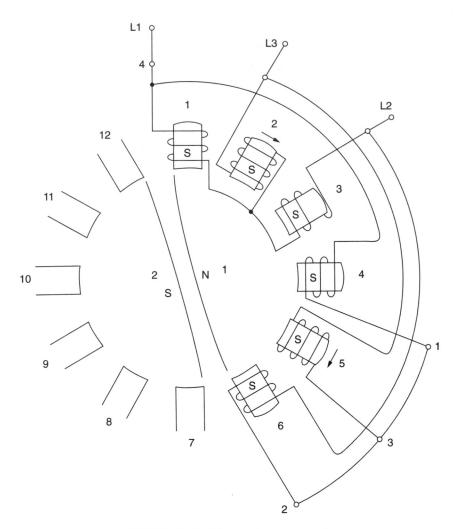

FIGURE 14-34   Three-phase, two-pole.

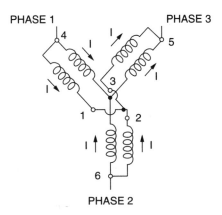

FIGURE 14-35   Schematic three-phase, high-speed with current direction (Point A, Figure 14-29).

> *WARNING*: Do not confuse the series and parallel connection of windings in the two-speed compressor for speed control with series and parallel connection of windings in standard motors for low and high voltage. The number of poles in a standard motor does not change when the winding connections are changed from series to parallel. The speed remains the same; only the voltage requirement changes.

VARIABLE-SPEED, THREE-PHASE COMPRESSOR

Another system of speed control for three-phase is rapidly being developed. The continuously variable speed control for three-phase compressor motors is now becoming available. The following is a general outline of a system that may be followed in the block diagram of Figure 14-36. Digital logic circuitry is used to develop the three-phase signals. The outputs of the three-phase logic block are three square waves of voltage that are 120° out of phase with each other, just as standard three-phase voltages are 120° out of phase. These waveforms are shown on the block diagram. The input control to the three-phase logic block is a variable-frequency pulse from a frequency generator.

A control input signal determines the frequency of the output. The control input could be a voltage from an electronic thermostat setting the compressor speed requirement. The level of the dc voltage from the thermostat indicates how far the temperature is from the selected temperature. The greater the deviation, the higher the output is. The higher the output is the higher the frequency output from the frequency generator. The higher the frequency input to the three-phase logic is, the higher the frequency output from the logic circuit. In other words, three-phase voltages of a variable frequency are available to drive the three-phase compressor motor. The speed of the compressor motor is determined by the frequency of the supply voltage.

$$rpm = \frac{2F \times 60}{N}$$

The three-phase logic circuit feeds the individual-phase wave shaper. The wave shaper provides pulses of current to the three-phase compressor motor. The motor is supplied with three-phase voltage and current and will run at the speed determined by the frequency. The higher the frequency is, the higher the speed.

When the compressor is operated at higher speeds, it can produce higher pressure. In a properly designed system, this will provide for more rapid cooling. If little cooling is required, the thermostat through the system will cause the compressor to operate at a lower speed, using less power.

SUMMARY

- Single-phase induction motors require a starting system.
- Shaded-pole motors rotate in the direction of the shaded pole.
- Split-phase motors have two sets of windings.

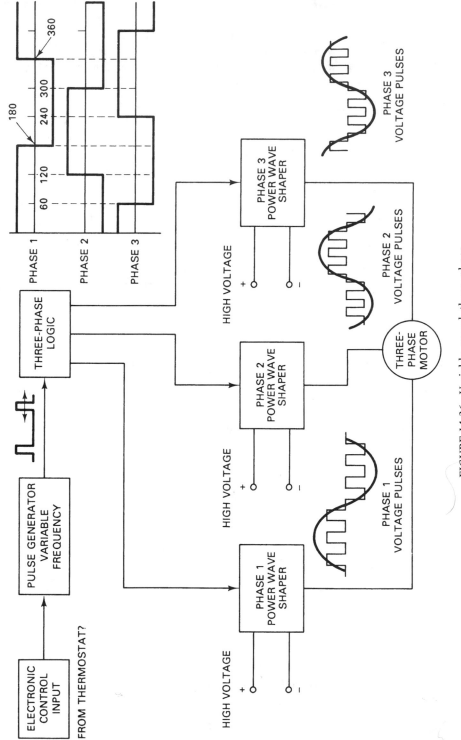

FIGURE 14-36 Variable-speed, three-phase.

start

- Capacitor motors provide higher torque. _(Capacitor run motors have lousy running torque)._
- Polyphase motors provide their own rotating magnetic fields. _(change leads to change direction)._
- Variable-speed motors provide for improved efficiency of air-conditioning systems. _(lower current use)_

-> system runs all the time (particulates, heating, humidity, ventilation all work better -- if they run constantly.

PRACTICAL EXPERIENCE

Required equipment Shaded-pole motor, compressor motor, ohmmeter.

Procedure

1. Observe the face of a shaded pole motor.
2. Locate the shaded pole from one side of the motor.
3. The rotation of the motor is from the unshaded section to the shaded section.
4. Determine the direction of motor rotation cw or ccw. _____
5. Observe the opposite side of the motor.
6. What is the rotation from this side?
7. Observe the three terminals of a compressor motor.
8. Mark the terminals 1, 2 and 3.
9. Measure the resistance terminal 1 to 2. _____ohms
10. Measure the resistance terminal 2 to 3. _____ohms
11. Measure the resistance terminal 3 to 1. _____ohms
12. The lowest resistance in steps 9, 10, and 11 indicates the run winding.
13. The second-to-lowest resistance is the start winding.

REVIEW QUESTIONS

1. Direct current (dc) motors are the most common type used to drive air-conditioning compressors.   T_____   F_____

2. In a shaded-pole motor, the shaft always rotates in the direction of the shaded pole.   T_____   F_____

3. The direction of rotation of a split-phase motor may be changed by interchanging the connections to the start winding.   T_____   F_____

4. Centrifugal switches are common in refrigerator sealed compressors.   T_____   F_____

5. Shaded-pole motors are more efficient than permanent split-capacitor motors.   T_____   F_____

6. Capacitor-start motors have high starting torque.   T_____   F_____

7. The run capacitor used with capacitor-start,
   capacitor-run motors is of a larger value than the
   start capacitor.   T_____   F_____

8. Three-phase motors use capacitors in their
   start circuit.   T_____   F_____

9. To reverse the direction of a three-phase motor,
   all three input wires must be changed.   T_____   F_____

10. Three-phase motors are sometimes started in the
    wye configuration and then switched to the delta
    configuration after the motor comes up to speed.   T_____   F_____

Motor-Starting Circuits

NOTE: READ! (March 5, 1998) ↑ I need to know This chapter! (for my career)

OBJECTIVES

Upon completion and study of this unit, you will be familiar with the starting devices, starting systems, and voltage control of electric motors. Devices covered are
- Series start (current) relay.
- Potential relay.
- Push-button.
- Relay-control, three-phase.

Different types of electric motors used in air-conditioning and refrigeration systems were covered in Unit 14. Some of these motors require special control for starting, whereas others are started simply by closing a switch. The air-conditioning and refrigeration technician should become familiar with the different types of motors and the circuits associated with them.

SHADED-POLE MOTORS

Shaded-pole motors are generally used in air-conditioning and refrigeration systems as a small fan motor, and as timing motors in large residential and commercial units. The shaded-pole motor does not require a special starting control. When power is supplied to the motor winding, the motor rotates. When power is removed from the motor winding, the motor stops but is ready to start again with the application of power. The starting feature of the motor, the shaded poles, is always in the circuit, ready to produce the required starting action.

PERMANENT SPLIT-CAPACITOR MOTORS

The permanent split-capacitor motor, like the shaded-pole motor, does not require special starting considerations. Simple switch devices, either manual or automatic, such as a thermostat, may be used to start this motor. When power is applied, the motor runs.

SPLIT-PHASE MOTOR (CURRENT RELAY)

The split-phase motor is started by the use of a centrifugal switch. The centrifugal switch removes the start (auxiliary) winding from the circuit when the motor shaft approaches operating speed. This is a practical system for starting the motor when it is used in open compressor-type applications. Centrifugal switch operation was covered in Unit 14.

One of the most common uses of a split-phase motor is to "drive" a compressor in small-capacity refrigerators and freezers. The split-phase motor in this application is hermetically sealed along with the compressor as a combined unit. In a sealed unit, a centrifugal switch starting system would not be practical. (See Figure 14-13.) Arcing at the switch contacts would contaminate the refrigerant-oil combination in the sealed system. Another problem is that oil on the switch contacts would increase contact resistance. For these reasons, centrifugal switches are not used in hermetically sealed compressors.

The most common type of starting control used with the small-capacity, hermetically sealed compressors is the series starting relay. This type of relay is known as a current relay (Figure 15-1). The coil of the current relay consists of a few turns of relatively large wire connected in series with the run winding of the motor. The contacts of the current relay are normally open and are connected in series with the start winding of the motor. A circuit diagram of a split-phase motor with a current relay is shown in Figure 15-2.

The relay coil is connected in series with the run winding of the motor. The open contacts of the relay are connected in series with the start winding of the motor. When the control switch is closed, 120 volts is applied to the run winding of the motor through the few turns of the relay coil. The start winding is not connected to the power source because the relay contacts are open. A high current flows in the run winding since the motor has not yet started. This same current flows through the relay coil (Figure 15-3).

FIGURE 15-1 Series start relays. (Courtesy of BET, Inc.)

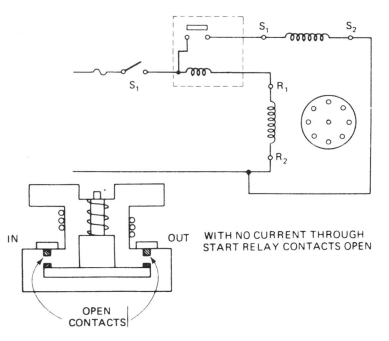

*FI*GURE 15-2   Series start relay with no current through start relay, contacts open.

With a high current flow in the coil, the relay energizes; the normally open (NO) contacts of the relay close. A circuit is completed through the relay contacts, and 120 volts is applied to the start winding of the motor. The motor starts to rotate and gains speed. When the motor approaches operating speed, the current flow in the run winding decreases toward the normal operating current (Figure 15-4). This reduction causes the relay to de-energize; the relay contacts then return to their normally open position, again removing the start winding from the circuit.

Current relays are commonly used with compressors up to $\frac{3}{4}$ horsepower capacity. Larger compressors commonly use the potential relay for starting control.

Many current-type starting relays operate with the weight of the relay core as a factor in opening the relay contacts. This type of relay must be mounted in the position prescribed in the manufacturer's specifications.

Current relays are designed to pick up (energize) and to drop out (de-energize) at specific current levels. The current relay used in a compressor motor circuit must be matched to the compressor motor. When it is necessary to replace a current relay, an exact replacement must be used.

CAPACITOR-START MOTORS (CURRENT RELAY)

Capacitor-start motors used in refrigeration and air-conditioning applications often use the same type of current relay that is used with a split-phase motor.

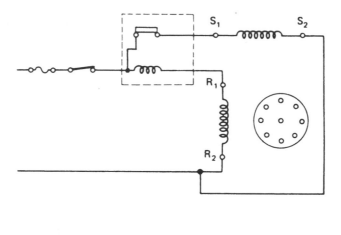

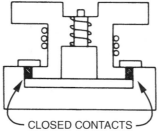

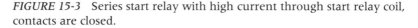

CLOSED CONTACTS

FIGURE 15-3   Series start relay with high current through start relay coil, contacts are closed.

A diagram of a capacitor-start motor circuit using a current relay is shown in Figure 15-5. When the switch is closed, high current flows through the coil of the current relay and the run winding of the compressor motor. The relay contacts close, connecting the start winding of the motor to the supply voltage through the start capacitor.

As the motor approaches operating speed, the run winding current decreases. When the current reaches a predetermined level, the relay de-energizes, removing the capacitor and the start winding from the circuit.

It is important to remember that the electrolytic capacitor used with capacitor-start motors is for intermittent use only. It is only in the circuit for a few seconds during the motor start period.

CAPACITOR-START, CAPACITOR-RUN MOTORS (POTENTIAL RELAY)

Capacitor-start, capacitor-run motors are usually provided with a potential relay as a starting control. This type of motor is generally used in air-conditioning and refrigeration applications from $\frac{3}{4}$ to 10 horsepower or larger.

The potential relay that is used in motor start control is similar in design and appearance to many other types of relays (Figure 15-6). The potential relay consists of a coil and a set of normally closed (NC) contacts. Potential relays are volt-

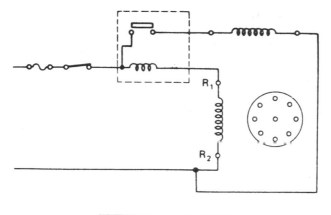

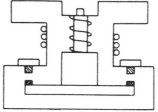

FIGURE 15-4   As motor comes up to speed, current draw decreases and the spring pressure plus gravity opens the contact.

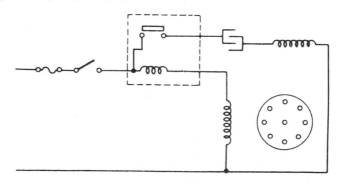

FIGURE 15-5   Series start relay connection, capacitor-start relay.

FIGURE 15-6   Potential relay. (Courtesy of BET, Inc.)

age sensitive; they are rated with pick-up (energize) and drop-out (de-energize) voltage. Exact replacement parts should always be used. Although many potential relays have a *sensitivity adjustment* associated with the contacts, adjustments should not be made in the field. If a potential relay does not function properly, it should be replaced. It is not a good practice to attempt adjustment or repair of relays.

The circuit of a capacitor-start, capacitor-run motor with potential relay start control is shown in Figure 15-7. The relay contacts are only used to connect and remove the *start* capacitor from the circuit. The run capacitor remains in the circuit at all times.

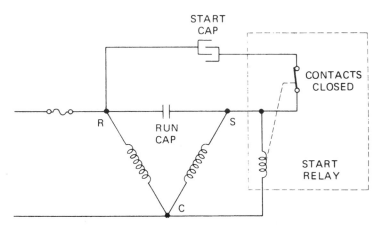

FIGURE 15-7   Potential relay connection to capacitor-start, capacitor-run compressor.

When the switch is closed, 220 volts is applied to the motor circuit. The run winding is connected directly to the 220-volt source. The start winding is connected to the 220-volt source through both the start and run capacitors. Electrically, the two capacitors are connected in parallel through the closed contacts of the potential relay.

The potential relay coil is connected in parallel with the start winding of the motor. The relay energizes when the voltage across the start winding rises to the predetermined level, depending on the specifications of the potential relay used. As voltage is applied to the motor, it immediately starts to rotate. High torque is present because of the high current in the start winding. The speed of the motor increases, and the back EMF across the start winding also increases. When the back EMF reaches the predetermined level, the relay energizes, opening the normally closed contacts. The start capacitor is then electrically removed from the circuit. The motor continues to rotate, with the run winding fed directly by the power source. The start winding is fed power with the run capacitor in series with the winding (Figure 15-8).

Potential relays, though manufactured by many companies, are often interchangeable. When replacement of a potential relay is necessary, an exact "specification" replacement must be used. The pick-up and drop-out voltage range of this type of relay is most important.

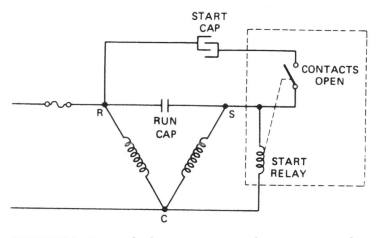

FIGURE 15-8 Potential relay contacts open when a motor speeds up.

The contacts of the potential relay may become pitted if high voltage stored in the start capacitor is discharged through the contacts as they close. To reduce the pitting problem, it is common practice to connect a 15,000- to 18,000-ohm, 2-watt bleed resistor across the terminals of the electrolytic capacitor. The resistor does not affect the operation of the capacitor or motor, but it does provide a discharge path for the capacitor when it is not in use.

PUSH-BUTTON, ON-OFF MOTOR CONTROL

A common control used in the start operation of motors is the push-button magnetic-type starter switch. One excellent feature of this system is that the control may be located away from the motor. A diagram of a push-button system is shown in Figure 15-9.

The motor is controlled by a magnetic contactor, K-1. A thermal overload protector is included in each line of the motor control. Should excessive current be drawn in either line, the thermal overload will overheat, causing the associated set of contacts to open, interrupting the circuit.

The normally open start button is pressed to start the motor. The current path is from line L_1, through the start button, now closed, through the normally closed stop button, through both thermal overload contacts to the bottom of the contactor coil. The top of the contactor coil is connected directly to line L_2. The contactor energizes, supplying power to the motor. The contacts between L_1 and T_1, when closed, complete the circuit, bypassing the start button. The contactor coil remains energized. The contactor coil de-energizes if the contacts of a thermal overload open or if the stop button is pressed.

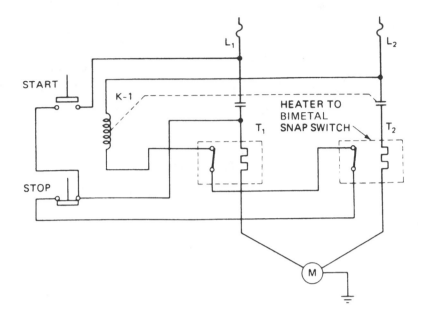

FIGURE 15-9 Push-button motor starter.

THREE-PHASE MOTOR

Three-phase motors are not started through the use of phase-shifting devices. A rotating magnetic field is present about the motor stator whenever three-phase voltage is applied to the windings. With larger units, a starting sequence is sometimes used to reduce initial starting current. During starting, the motor is electrically connected in the wye configuration; it is electrically reconnected in the delta configuration when operating speed is approached. Through the use of such connections, the voltage across the motor windings is reduced during the starting sequence, but is at the full level during operating speeds.

In Figure 15-10a, the three-phase motor connection for start-up is in the wye configuration. The voltage across each winding is 254 volts. The source voltage is 440 volts. Full-coil voltage operation with the delta connection is shown in Figure 15-10b. The operation of the contactor or relay used to switch from wye to delta operation may be controlled by a time-delay mechanism. When power is first applied to the motor, the relay is de-energized, and the circuit is wye-connected. After a fixed time delay, during which the motor picks up speed, the relay energizes, changing the motor to a delta connection. The voltage across the motor windings is 440 volts.

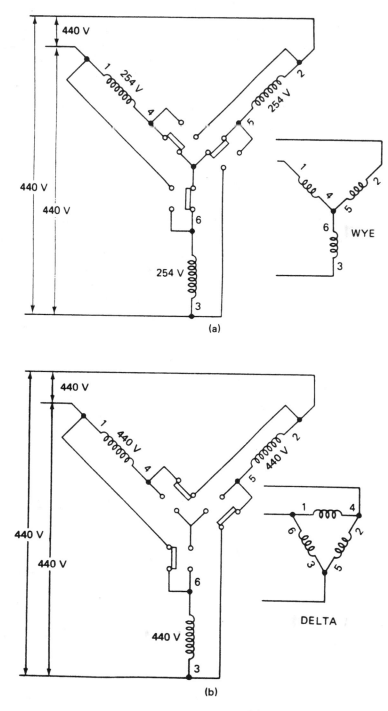

FIGURE 15-10 Wye start, delta run.

Single Phasing of Three-Phase Motors

When a three-phase motor is connected to a three-phase power source, it starts and eventually reaches normal operating speed. If, after the motor is started, one of the three input power connections becomes open, single-phase power is supplied to the motor. The motor continues to operate as a single-phase motor; however, excessively high current is drawn, and the motor overheats. If the motor is allowed to continue to operate in a single phase, permanent damage to the motor may result. Thermal protective devices are connected in most three-phase motor circuits to electrically disconnect the motor if it becomes overheated.

LOW/HIGH-VOLTAGE OPERATION

Many motors are wound with coils in pairs. With such motors, it is possible to connect them electrically to operate on either high (240) or low (120) voltage. In Figure 15-11a, each coil is designed to function with 120 volts applied. In Figure 15-11b, the coils are connected in parallel for operation from a 120-volt source. So connected, each coil has 120 volts applied to it.

In Figure 15-11c, the coils are connected in series for operation from a 240-volt source. As shown, each coil has 110 volts across it. For a better understanding of this arrangement, review Units 4 and 5.

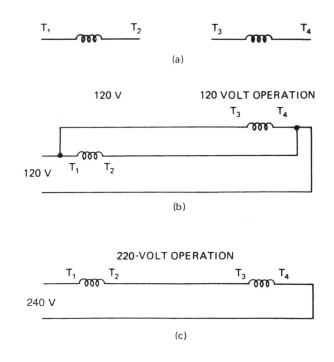

FIGURE 15-11 Examples of
120 V/ 240 V operation.

Single-Phase Motors: Low and High Voltage

Many dual-voltage, single-phase motors have two coils for the run winding and two coils for the start winding. Winding connections for low- and high-voltage operation are shown in Figure 15-12a and b.

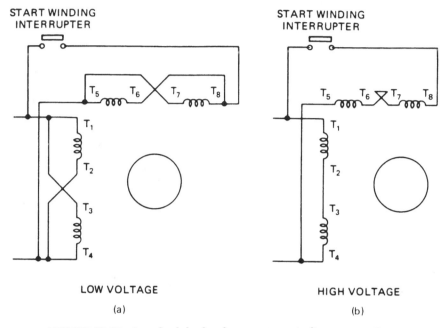

FIGURE 15-12   Standard dual-voltage motor winding connections.

Other dual-voltage, single-phase motors have two coils for the run winding and a single coil for the start winding. Winding connections for low- and high-voltage operation are shown in Figure 15-13a and b. When connected for low-voltage operation, the two coils of the run windings (T_1-T_2 and T_3-T_4) are connected in parallel across the 120-volt power source. The start winding T_5-T_6 is also connected directly across the 120-volt power source.

For operation at higher voltage, the run windings are connected in series. The start winding is connected to the junction of T_2 and T_3. Slightly less than one-half of the source voltage appears across the start winding during motor start, as shown in Figure 15-14a. After the motor comes up to operating speed, the switch device removes the start winding from the circuit. A schematic diagram of the circuit is shown in Figure 15-14b.

Three-Phase Motors: Low and High Voltage

The connection of windings for operation on high or low voltage with three-phase motors is a series or parallel consideration. In Figure 15-15, the motor is connected for operation at the low voltage. The winding pairs of each phase are connected in parallel. For high-voltage operation, the winding pairs for each phase are connected in series. This configuration is shown in Figure 15-16.

Control of motors is covered further in Units 16 and 18.

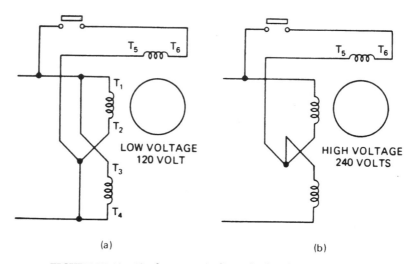

FIGURE 15-13   Single start winding, dual-voltage operation.

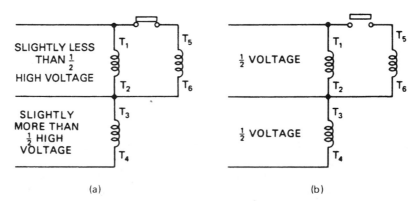

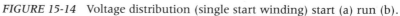

FIGURE 15-14   Voltage distribution (single start winding) start (a) run (b).

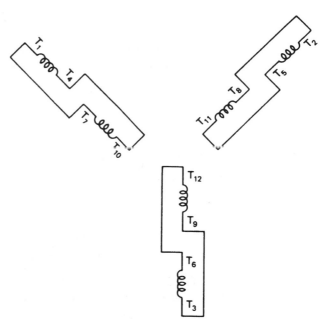

FIGURE 15-15 Three-phase motor, low-voltage parallel connection.

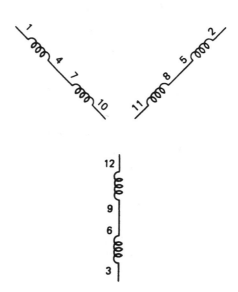

FIGURE 15-16 Three-phase motor, high-voltage series connection.

SUMMARY

- Most AC electric motors require a special starting system for proper operation.
- The nameplate information of the motor is important and should be followed when making motor connections.

PRACTICAL EXPERIENCE

Required equipment   Ohmmeter, series start relay (current relay), potential relay.

Procedure

1. Connect the ohmmeter across the contact terminals of the series start relay.
2. Does the meter indicate an open or short terminal?
3. With the meter connected to the relay, turn the relay over.
4. Did the circuit through the relay change from open to short or vice versa?
5. Connect the meter across terminals 2 and 5 of the potential relay.
6. Measure and record the resistance of the relay coil. _____ ohms
7. Connect the ohmmeter across terminals 1 and 2 of the relay.
8. Are the contacts open or shorted?

Required equipment   Dual voltage single-phase or three-phase motors.

Procedure

1. Observe the nameplate on the motor.
2. Locate wiring information for low-voltage operation.
3. Properly connect the motor for low-voltage operation.

Question   Are the coil sets connected in series or parallel for low-voltage operation?

4. Have the motor checked by supervisory personnel for proper connection.
5. Connect the motor to the proper voltage through suitable control devices.
6. Turn the motor on and observe the motor rotation.
7. Disconnect the motor from power.
8. Make motor winding connections required to reverse the direction of the motor.
9. Repeat steps 5, 6, and 7.
10. If time permits, repeat this experiment with the motor connected for high-voltage operation.

REVIEW QUESTIONS

1. The shaded-pole motor is used in low-power
   applications.   T_____   F_____

2. A permanent split-capacitor motor requires a special starting device, called a centrifugal switch.   T_____   F_____

3. The starting device used in split-phase motors in hermetically sealed compressors is the centrifugal switch.   T_____   F_____

4. The potential relay coil is connected in series with the run winding of the motor.   T_____   F_____

5. The current relay has contacts that are open when the relay is de-energized.   T_____   F_____

6. The potential relay has contacts that open when the motor approaches operating speed.   T_____   F_____

7. It is not good practice to make adjustments or repairs on relays.   T_____   F_____

8. All current relays may be operated in any position.   T_____   F_____

9. Many high-horsepower motors in air-conditioning systems use potential relays.   T_____   F_____

10. Current or series relays are used in high-horsepower applications.   T_____   F_____

Unit **16**

Control Devices

OBJECTIVES

Upon completion and review of this unit, you will understand
- Power switches.
- Fuses and circuit breakers.
- Contactors and relays.
- Solenoids.
- Thermostats.
- Pressure controls.

Many different types of control devices are used in air-conditioning and refrigeration circuits. These controls may be classified into four main divisions:

1. Personal protection
2. Circuit protection
3. Operation control
4. Operation protection

The devices in each classification are covered in this unit.

PERSONAL PROTECTION

Perhaps the most important control associated with electrical devices is the main power interrupter. With domestic refrigerators or through-the-wall air-conditioning units, the power interrupter may simply be the wall plug. The technician may "pull the plug" and work on the unit knowing that power to the unit has been removed.

In larger systems in which power is supplied from a distant power panel, a power disconnect switch is usually located near the condensing unit. The technician may use this switch to interrupt power for maintenance or inspection. Whenever the power is interrupted by the technician for these purposes, the switch should be *locked* in the open position (Figure 16-1). *Use an actual lock.* This is of particular importance if the switch is not located in the immediate area in which the technician is working.

FIGURE 16-1 Main power switch locked open. (Courtesy of BET, Inc.)

(handwritten note: ① Be certain: check power !!)

Many technicians have received serious electrical shock while working on units when power was thought to have been disconnected. Be certain: use the lock. *(handwritten ②)*

A common safety switch includes fuses for overload protection in the circuit. The safety switch may be used to disconnect power from a unit requiring maintenance or inspection. It is sometimes necessary to replace the fuses in a safety switch. Whenever a fuse is to be replaced in a system with which you are not familiar, a voltage check should be made at the fuses. Safety switches have been wired in backward, creating a dangerous situation. The proper way to wire in a safety switch is shown in Figure 16-2a. When the safety switch is open, the fuses are isolated from power.

The incorrect connection is shown in Figure 16-2b. With this connection, when the switch is open, power is still connected to the fuses. An attempt to replace a fuse might result in a dangerous shock.

Before attempting to replace any fuse, check for voltage between the ends of the fuses as shown in Figure 16-3 and also from each fuse terminal and ground.

CIRCUIT PROTECTION

The most common type of electrical circuit protective device is the simple fuse. A *fuse* is a one-time overcurrent protective device. When blown, it must be replaced. The electrical portions of a fuse are the fuse link and the input and output connection terminals. Whenever current above the rated-value limit of the fuse is drawn (through the fuse), sufficient heat develops to melt the fusable link; this permanently opens the circuit.

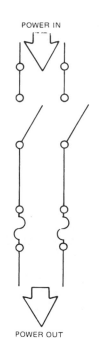

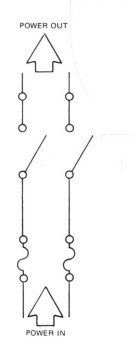

FIGURE 16-2(a)   Correct connection of safety switch.

FIGURE 16-2(b)   Incorrect connection of safety switch.

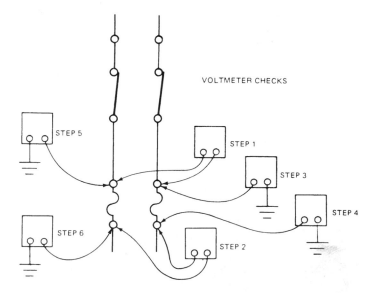

FIGURE 16-3   Voltmeter check of fuse circuit.

Fuses are available in many physical sizes and shapes, as shown in Figure 16-4. Many applications require the use of *time-delay* fuses. A time-delay fuse is one in which current beyond its rated value may be drawn for a limited time without a burnout. This fuse allows for the brief start-up of motors (high starting current), but still protects in case of continuous overload or short circuits.

FIGURE 16-4   Fuses. (Courtesy of BET, Inc.)

Many cartridge fuses, particularly the larger, amperage-rated, knife-blade type, are constructed in a manner to provide a means of replacing a burned-out element. Fuse elements of the same ampere rating should be used for replacement. Whenever a fuse with a replacement element is to be installed, it should be disassembled first to ensure that the correct ampere-rated element is in place.

Edison-base fuses are often found in older installations in applications requiring up to 30 amperes. The Edison-base fuse should not be used in newer installations. The S-type fuse with an S-adaptor should be used in new installations. The problem with the Edison-base fuse is that fuses of any value can be installed in any fuse socket. With the S-type base, only the prescribed amperage-value fuse fits in the base. The base usually includes a locking device to ensure that the base cannot be removed from the socket and replaced with a larger-amperage-value S-type base.

HRC fuses

Circuit Breakers

Circuit breakers are available in two general types, magnetic and heat overload. Although the popularity of the magnetic circuit breaker is increasing, the number of circuit breakers in use that function in this manner has not yet approached the number that employ a heat-operated overload (heat proportional to current).

Both types function in a similar manner in a portion of their operation. In the ON (closed) position, the switch mechanism is spring loaded and locked. The control (release) mechanism functions with the lock portion of the circuit breaker. When the lock is released, the spring mechanism forces the circuit breaker switch to the OFF (open) circuit position. It remains in this position until it is manually reset.

→ MCP (motor circuit protector) — designed specifically for motor circuits

Magnetic Circuit Breakers   In the magnetic circuit breaker, the line current passes through an electromagnet (Figure 16-5). The electromagnet controls the lock mechanism, which holds the circuit breaker (switch) in the ON position.

handle motor inrush current without blowing

When an overload (high) current passes through the electromagnet coil, the magnetic field is of a magnitude sufficient to release the lock mechanism. The spring-loaded switch snaps to the OFF position.

Magnetic circuit breakers are fast acting. The circuit breaker may be manually reset to the ON position as soon as the problem causing the overload has been corrected.

magnetic trip coil
— can handle

FIGURE 16-5 Magnetic circuit breaker. (Courtesy of BET, Inc.)

Heat-Overload Circuit Breakers   The most common type of circuit breaker operates on the principle of heat overload reacting on a sensitive bimetal strip. The bimetal strip controls the lock-in mechanism (Figure 16-6). Line current enters the circuit breaker at the top and passes through the switch contacts, the bimetal strip, and the output terminal. The bimetal strip and associated levers and springs mechanically control the lock-in mechanism. In the case of overload current, the bimetal strip heats and physically changes shape.

FIGURE 16-6 Heat-overload circuit breaker. (Courtesy of BET, Inc.)

As the bimetal trip changes shape, it engages the trip mechanism and releases the lock mechanism. The locking arm is released, the compression spring is released, and the expanded spring pulls the contacts open.

Before the circuit breaker can be reset, the overload condition must be removed from the line. There is an extra delay because the bimetal strip must cool down before the lock and trip mechanism may be reengaged.

The lock mechanism may be interconnected to more than one set of control contacts. Figure 16-7 is a photograph of a 40-ampere circuit breaker of the heater type; if an overload occurs in either line, the interlocking mechanism opens both sets of contacts. The same type of mechanism may open three sets of main contacts for three-phase applications.

FIGURE 16-7 Interlocked circuit breakers. (Courtesy of BET, Inc.)

OPERATION CONTROLS

The controls associated with motors in air-conditioning and refrigeration applications are, for the most part, standardized. As discussed in Units 14 and 15, for low-horsepower motor applications, simple switches are used to turn them off and on.

As the horsepower rating is increased and/or the type of motor is changed, the number and types of control devices associated with the compressor motor increase.

Primary Controls

The primary controls associated with compressor motors are those that have a direct effect on the operation of the motor. In most applications, the controls consist of the following:

1. Main contactor or power relay
2. Start relay
3. Overload (thermal)

Contactor/Relay

Figure 16-8a represents a 24-volt ac contactor. In its de-energized state, the spring (1) holds the bar (2) in the up position. There is a complete circuit from terminal A through the contactor bar to terminal B. This would be shown schematically as a normally closed set of contacts A and B (Figure 16-8c). The normally open set of contacts are shown as C and D in Figure 16-8d. When 24 volts ac is applied to the contactor coil, the coil energizes. The magnetic field pulls the armature (3) to the center of the coil. The bar disengages contacts A and B and mates contacts C and D. When the 24 volts is removed from the coil, it will de-energize. The spring will push the bar up, reengaging A with B. Other bar and contact combinations may be attached to the armature.

Figure 16-8b represents a 24-volt ac relay. One set of contacts is shown. The armature (1) rotates about the pivot (2) point. The spring (3) holds the left end of the armature down when the coil is not energized. A wire connects terminal A to the armature, which, when the coil is de-energized, makes a connection of A to B. When the coil is energized by the application of 24 volts ac, the magnetism of the coil attracts the relay armature. The right end of the armature moves down, rotating about the pivot point. The connection from A to B is broken, while the connection from A to C is made. The schematic symbol for the relay contacts is shown as Figure 16-8e. A relay may have many sets of contacts.

Main Contactor   The main contactor (power relay) is used to control the application of high- (line-) voltage electric power to the motor circuits. The contactor is usually controlled by a low-voltage coil. When the coil is energized, contacts are mated, and high voltage becomes available to the circuits as required. Figure 16-9 is a photograph of a power relay. The circuit diagrams of the device are also given.

Motor-Start Relays   The motor-start relays used in air-conditioning and refrigeration circuits are of the current type or the potential type. Both relays were covered in Unit 15.

Most of the secondary controls that affect the operation of an air-conditioning or refrigeration system are connected so as to control the application of low voltage to the coil of the main contactor.

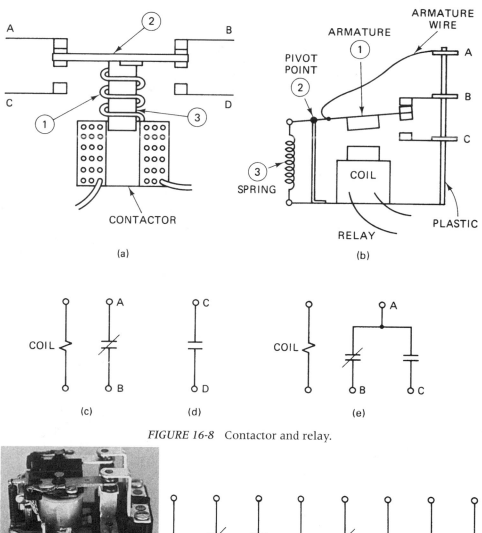

FIGURE 16-8   Contactor and relay.

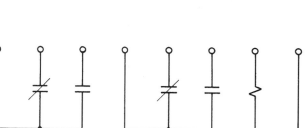

FIGURE 16-9   Power relay.

Thermal Overload   A thermal overload is used with most compressors in present-day operations. The overload opens to disconnect power to the compressor motor whenever excess heat causes the internal bimetal disc to snap open.

   The thermal overload is placed in physical contact with the compressor case to sense heat from within the compressor (Figure 16-10). If the compressor becomes too hot for safe operation, the bimetal disc bends and snaps open, removing electric power from the motor common terminal.

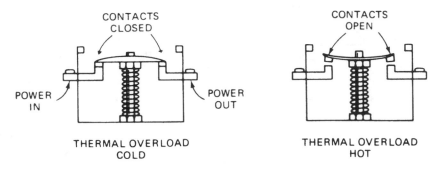

FIGURE 16-10   Thermal overload—normal(left) and overheated(right). (Courtesy of BET, Inc.)

In some older systems, a three-terminal thermal overload is used (Figure 16-11). With this type of overload, the temperature of the compressor, as well as excessive starting current, causes the overload to open. The start winding is connected in series with the overload heater. Excessive current or start current drawn for too long a time heats up the overload, causing the bimetal disc to snap open. The two- and three-terminal overloads are connected in different electrical positions (Figure 16-12).

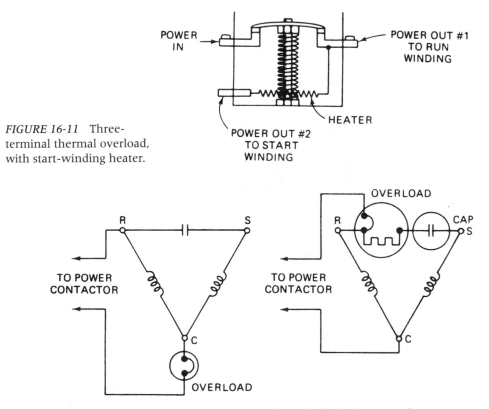

FIGURE 16-11   Three-terminal thermal overload, with start-winding heater.

FIGURE 16-12   Two- and three-terminal overload circuit connections.

Multiple-Contact Relays

Multicontrol (contact) relays are used for a special application in air-conditioning and refrigeration electrical circuits. The purpose of a multiple-contact relay is to provide for an automatic multiswitching sequence. Figure 16-13 is a photograph of a multiple-contact relay. The schematic diagram of the relay is also shown.

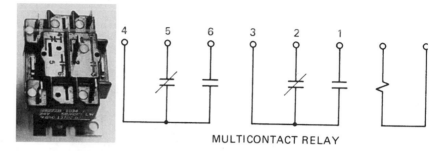

FIGURE 16-13   Multicontact relay. (Courtesy of BET, Inc.)

Thermal Relay (Time Delay)

Thermal time delay has been used in system control for a number of years. The relay consists of a heater, a bimetal strip, and a set of contacts (Figure 16-14). When power is applied to control terminals 1 and 2, the internal heater warms up. The bimetal strip bends, completing the circuit between terminals 4 and 3. Thermal time-delay relays are available with variable delays between control-power application and contact closure. A photograph of a time-delay relay is shown in Figure 16-15.

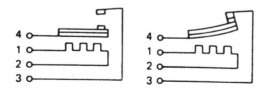

FIGURE 16-14   Bimetal time-delay relay.

FIGURE 16-15   Bimetal time delay. (Courtesy of BET, Inc.)

Solenoids

Solenoids are used in the air-conditioning and refrigeration industry to provide a means of electrical control of electromechanical valves. The valves may be in refrigerant lines to control reverse cycle or defrost operations or in water lines to control water towers, water-cooled condenser units, or chilled-water air-conditioning systems.

A sketch of a solenoid valve without power supplies is shown in Figure 16-16. The spring-loaded valve arm holds the valve needle tight against the valve seat, obstructing flow from the input connection to the output connection. When power is applied to the solenoid coil, the magnetic attraction of the coil overcomes the spring and pulls the solenoid armature with the valve needle up into the solenoid core. The valve needle is upseated from the valve seat, and a flow path is opened from the input connection to the output connection.

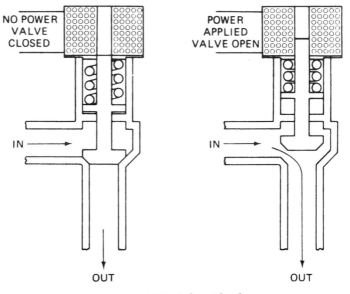

FIGURE 16-16   Solenoid valve.

Caution must always be taken in the connection of input and output lines. When the lines are properly connected, the pressure from the input line tends to keep the valve closed (Figure 16-17). If the lines are connected in reverse, the pressure from the input line pushes against the valve-head surface, tending to open the valve.

Two-way solenoid valves are also available and are used in some systems (Figure 16-18). When the solenoid coil is de-energized (Figure 16-18a), the spring-loaded valve head is held down against the output 2 valve surface. A flow path is open from the input past the valve arm to output 1.

When the solenoid is energized (Figure 16-18b), the valve arm is pulled up. The valve head is positioned against the top valve surface, closing off output 1. As the valve moves up, the output 2 path for flow is opened.

Thermostats

Thermostats are made up of a temperature-sensing device and a control mechanism. Three main types of temperature-sensing devices are presently being used in the industry:

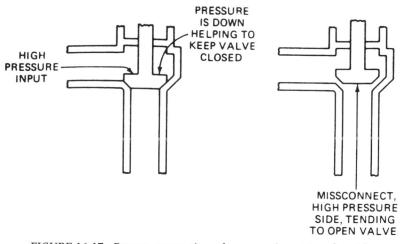

FIGURE 16-17 Proper connection of pressure input to solenoid.

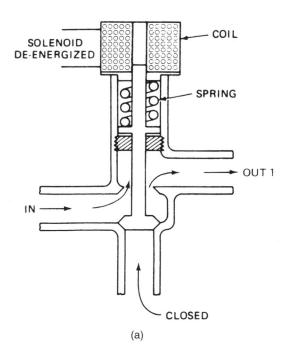

FIGURE 16-18(a) Two-way solenoid valve, solenoid de-energized.

(a)

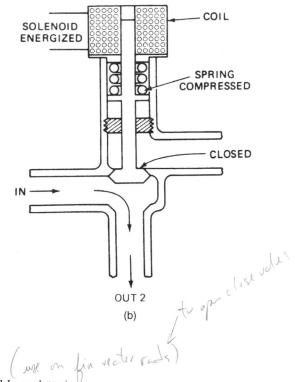

FIGURE 16-18(b)   Two-way solenoid valve, solenoid energized.

(b)

to open close valves

1. Bimetal strips
2. Temperature-sensitive bellows   *(use on fin vector rads)*
3. Thermistor (temperature-variable resistor)

The bimetal strip thermostat is the most common for room or space temperature-control applications. The bimetal strip is usually formed into a coil to provide greater surface area and therefore greater accuracy in movement (Figure 16-19). This thermostat uses contacts connected to the bimetal strip for electrical control. It is designed for low-voltage operation and controls the operation of a compressor contactor (power relay).

FIGURE 16-19   Simple bimetal
thermostat. (Courtesy of BET,
Inc.)

Often, thermostats are available with several different subbases (Figure 16-20). A greater number of control applications are thereby possible with a single type of sensing unit (thermostat). Changes in circuit configuration are obtained by using a standard sensing unit and the subbase that provides the needed circuitry.

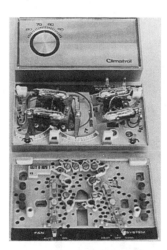

FIGURE 16-20 Two-stage heat, two-stage cool thermostat with subbase. (Courtesy of BET, Inc.)

When a thermostat is set at a specific temperature, the actual temperature at which it functions varies. To limit and control the variation, heat and cold anticipators are added to most thermostats. While the cold anticipator is fixed, the heat anticipator is often an adjustable element.

In a bimetal thermostat, as the room cools, the bimetal elements bends and makes a connection between the two contacts. The completed circuit provides for turn-on of the heating unit (Figure 16-21). The room heats as warm air is supplied to it. As the temperature of the room (and the thermostat) rises, the bimetal heats. At a predetermined high, the contacts open, shutting off the heater unit. If the selected room temperature is 74°F (23°C), the room temperature may vary between 72°F and 76°F (22°C and 24°C) with a bimetal thermostat. Figure 16-22 shows a graph of room temperatures.

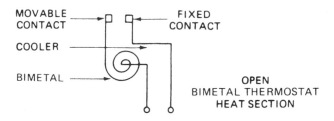

FIGURE 16-21 Bimetal thermostat.

Heat Anticipator The addition of a heat anticipator to the circuit helps to limit the temperature excursions. The heat anticipator is a resistance heater element connected in series with the thermostat line to the heat contactor coil (Figure 16-23). When the thermostat contacts are closed, current flows through the resistance heating element in the thermostat. The heat generated by the anticipator causes the thermostat to open shortly before the desired room temperature is reached. After the room heating unit is shut down, heat already generated in the unit continues to warm the room for a short period of time.

After the contacts of the thermostat open, current flow through the resistive heater in the thermostat stops. The bimetal cools down along with the room temperature, and the contacts of the thermostat again close at 74°F (23°C). Compare the

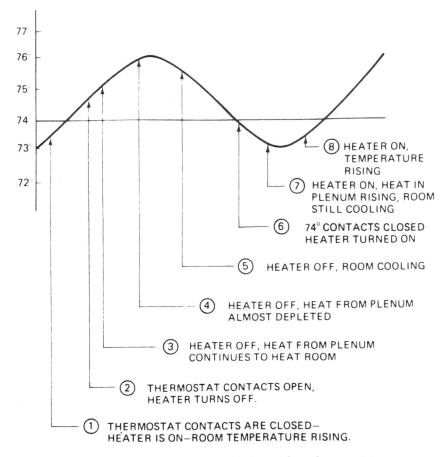

FIGURE 16-22 Temperature variation without heat anticipator.

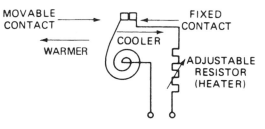

FIGURE 16-23 Bimetal thermostat
with heat anticipator.

graph in Figure 16-24 of room temperature when a heat anticipator is used with
the graph in Figure 16–22 of room temperature when no anticipator is employed.

The heat anticipator causes the furnace to cycle on and off more frequently,
while keeping the room temperature closer to the selected temperature.

Cold Anticipator The cold anticipator works on the same principle as the heat
anticipator. The connection of the heater resistor for the cold anticipator is parallel with the cooling contacts of the thermostat (Figure 16-25). The cold anticipator usually is not an adjustable resistor.

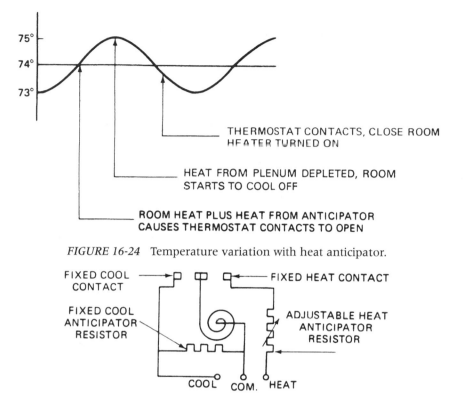

FIGURE 16-24   Temperature variation with heat anticipator.

FIGURE 16-25   Addition of cool anticipator.

When cool is selected, the cool anticipator resistor heats the thermostat bimetal during periods in which the cool contacts are open. This causes the contacts to close before the room temperature rises past the selected room temperature. When the air conditioner is turned on, the thermostat internal heater is shorted out by the cool contacts. No heat is developed in the cold anticipator resistor while the air conditioner is operating.

Repairs Field repair of thermostats by the service technician is not recommended. For service or replacement, the manufacturer or manufacturer's representative should be contacted. The electrical contacts of the thermostat may be cleaned when necessary. To clean contacts, a piece of clear, unprinted paper should be drawn through the closed contacts. Do not use sandpaper or a point file on the contacts.

Mercury Bulb Contacts Open contacts used in thermostats are a constant source of trouble since they are exposed to the air, and dirt and lint can get between them. Corrosion may also form on the contact surfaces, thereby causing poor connection capabilities.

Mercury switches are used in quality thermostats to overcome the deficiencies of the open contact system. A mercury bulb switch is connected to one end of the bimetal arm. As the temperature around the bimetal element changes, its position changes. This change moves the mercury switch, causing the contact to be closed or opened, as required. Both two- and three-terminal mercury bulb switches are used in thermostats (Figure 16-26).

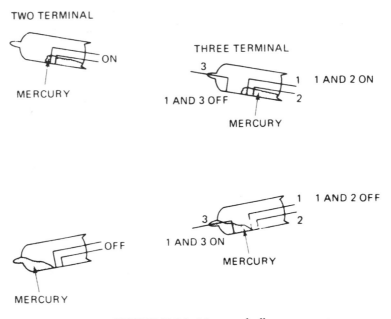

FIGURE 16-26   Mercury bulbs.

Some thermostats are designed for use with systems in which two stages of heating and/or two stages of cooling are required (Figure 16-27). Generally, the second stage is needed only under extreme temperature conditions. The mercury switches are placed in such a manner that stage 1 heat always comes on at a slightly higher temperature than stage 2 heat. In the cooling section, stage 1 cool comes on at a slightly lower temperature than stage 2 cool. This system provides for the turn-on of a single stage of heating or cooling when the temperature varies slightly from the selected temperature. When more extreme variations from the selected temperature occur, the thermostat calls for the operation of both stages of heating or cooling. Figure 16-28 shows thermostat positions for different temperature variations.

Although most thermostats are designed for low-voltage (24-volt) operation, some are on line voltages of 120 or 220 volts. In line-voltage types, the switch contacts of the thermostat open (or close) the high-voltage circuits.

Most air-conditioning applications recommend the use of low voltage in control circuits. The thermostat interrupts the application of 24 volts to control devices.

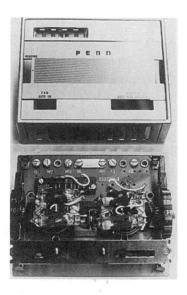

FIGURE 16-27 Mercury bulb thermostat. (Courtesy of BET, Inc.)

Microprocessor-Controlled Thermostats

The development of air-conditioning control circuits has kept abreast with the development of digital and microprocessor circuits in electronics. An example of a modern microprocessor-controlled thermostat is shown in Figure 16-29, the Honeywell T8300 thermostat and Q6300 subbase.

A portion of Honeywell publication 68-0025 is included on page 216 to provide a description of the features of the system, specifications, and operation of the unit. The wiring diagram of a two-heat, two-cool heat pump system (Figure 16-30) is included to show one of the connection systems possible with this thermostat. Notice that the circuit requirements external to the thermostat are the same as would be required with any control system. The sensing, computing, and controlling commands all come from the thermostat.

Heat-Limit Thermostats

Safety devices called limit controls are used with most heating systems. Limit controls (switches) are used in different areas of the heating system. The limit switch is designed to open if the temperature at the sensing point is above a predetermined limit.

In the furnace, a limit switch is located in the plenum chamber (Figure 16-31). If the heat in the plenum rises to a dangerous level, the limit thermostat opens, removing power from the electrically operated controls. The limit switch may be connected in either the 24-volt control circuit or the 120-volt line circuit.

Limit switches, also called high-limit controls, are connected in oil-burner-heating and strip-heating systems in much the same manner. Excessive temperature in the plenum chamber causes the limit switch to open, removing electrical power from the oil burner control valves or the heat strip contactors.

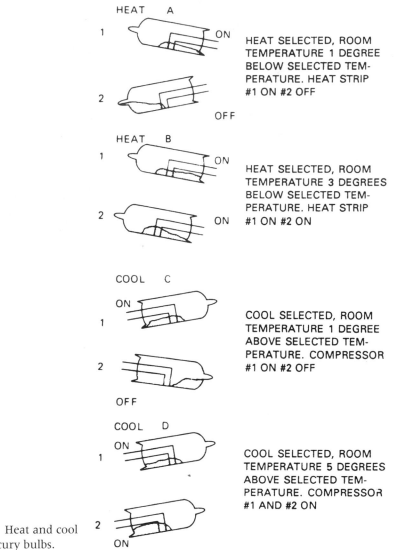

FIGURE 16-28 Heat and cool positions, mercury bulbs.

Bimetal Snap Limits (Trade name KLIXON)

Electric heat-strip units used with modern air-conditioning systems usually have a bimetal limit switch built into the heat strip assembly. If, for any reason, the temperature at the heat-strip unit rises above a safe limit, the bimetal switch snaps open, removing power from the heat strip. The bimetal limit switch is connected in series with the heating element (Figure 16-32).

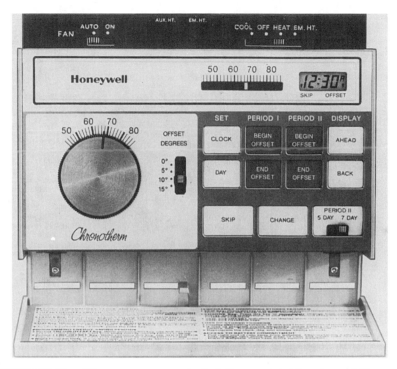

FIGURE 16-29a Microprocessor-controlled thermostat. (Courtesy of Honeywell, Inc.)

As an added protection, a special fuse is included as an integral part of the heat strip. Should the bimetal limit switch fail to open at the high-temperature limit, the fuse link will melt, removing power from the heat strip.

Pressure-Control Switches

Pressure-control switches are used as safety and/or temperature-sensing devices. The heat-limit switch shown in Figure 16-33 is a pressure switch that is sensitive to temperature change.

Pressure switches, regardless of type, have a common element, the bellows. The bellows is an expanding-contracting element that operates the switch or switches. The heat-limit switch shown in Figure 16-33 is an example of a bellows-operated switch.

The switches in Figure 16-33 are both normally open. They could, however, have both been normally closed, or one with the other. Switch functions NO or NC or any combination are found as required by the application of the device.

Temperature-Sensitive Pressure Switch   When the pressure input (capillary) tube to a pressure switch is closed at the end and is filled with an inert gas or liquid, it becomes sensitive to temperature (Figure 16-34). Often a remote bulb is soldered to the input tube to hold a larger supply of inert material in the miniature closed system. As the temperature rises, the inert material expands, causing the

T8300 SPECIFICATIONS

MODELS: T8300A Microelectronic Chronotherm Heat Pump Thermostat with Intelligent Recovery*. Provides two stages heat and one stage cool. Used in heat pump systems with gas, oil, or electric auxiliary heat.

T8300B Microelectronic Chronotherm Heat Pump Thermostat with Intelligent Recovery*. Provides two stages heat and two stages cool. Used in heat pump systems with a two-speed or dual compressor.

ELECTRICAL RATING: 20 to 30 Vac, 50/60 Hz power supply. Voltage must be full wave ac or device will not function properly. Both sides of transformer secondary are required at the subbase.

CYCLES PER HOUR:
   Heating—Nonadjustable. Factory-set at 3 cph for 1st stage and 2nd stage heat. 6 cph for 2nd stage heat available as a factory option.
   Cooling—Nonadjustable. Factory-set at 3 cph.

TEMPERATURE SETTING RANGE: 42 F to 88 F [7 C to 33 C].

THERMOMETER RANGE: 42 F to 88 F [7 C to 33 C].

ENERGY SAVINGS SETTINGS: 0 F, 5 F, 10 F, and 15 F [0 C, 3 C, 6 C, and 9 C] offset.

AMBIENT TEMPERATURE RANGE:
   Shipping—Minus 30 F to plus 150 F [minus 34 C to plus 66 C].
   Operating—plus 40 F to plus 110 F [4 C to 43 C].

HUMIDITY RANGE: 5 to 90 percent relative humidity.

DISPLAY: Liquid crystal display shows time of day with ±1 minute per month accuracy, and skip and offset indicators as appropriate. During programming and upon demand, it shows program times and day of week.

KEYBOARD: Sealed membrane programming keyboard requires pushing two keys simultaneously to minimize inadvertent program changes.

ENERGY SAVINGS PROGRAM: 1 or 2 energy savings time periods can be set. Five day/7-day switch allows user to choose whether Period II runs 5 days on and 2 consecutive days off, or on all 7 days. Skip and change keys allow user to temporarily override a scheduled program without changing the permanent programming. Programming is retained through power outage of 30 to 60 minutes; clock stops during outage and must be reset.

DIMENSIONS: See Fig. 1.

MOUNTING: T8300 mounts directly on Q6300 Subbase.

FINISH: Beige case with pewter-finished front scaleplate. Padded, beige cover conceals programming keys, thermostat setting dial, and slide switches on thermostat.

REQUIRED ACCESSORY: Q6300 Heating-Cooling Subbase.

Q6300 SPECIFICATIONS

MODEL: Q6300 Heating-Cooling Subbase. Provides COOL-OFF-HEAT-EMERGENCY HEAT system switching and AUTO-ON fan switching.
   Q6300A—Use with T8300A.
   Q6300B—Use with T8300B.

ELECTRICAL RATINGS:
   Power Supply—20 to 30 Vac, 50/60 Hz.
   Output Voltage—12 Vdc with T8300 connected. 35 Vdc max. with no load.
   Relay Contacts—1.2 A running; 3.0 A inrush on auxiliary heat contacts, 5.0 A inrush on compressor and fan contacts. Max. 30 Vac.
   Switch Contacts—1.5 A running and 7.5 A locked rotor. Max. 30 Vac.

WIRING CONNECTIONS: Screw terminals designed for either straight or wrapped connection.

TEMPERATURE RANGE:
   Shipping—Minus 30 F to plus 150 F [minus 34 C to plus 66 C].
   Operating—Plus 40 F to plus 110 F [4 C to 43 C].

HIGH LIMIT: Turns off relays when room temperature reaches 99 F [37 C] nominal.

DIMENSIONS: 6-7/16 in. wide, 4-1/2 wide, 9/16 in. deep [163.3 mm wide, 113.5 mm high, 114.5 mm deep].

MOUNTING: Mounts on wall or horizontal outlet box with screws provided. Two plastic anchors included for wallboard or plaster mounting.

OPERATING MODE INDICATION:
   Q6300A—LED's labeled AUX. HT. and EM. HT. light when thermostat calls for auxiliary heat or emergency heat operation.
   Q6300B—LED labeled EM. HT. lights when thermostat calls for emergency heat operation.

FINISH: Dark brown.

ACCESSORY: 196393A Cover Plate Assembly for mounting on vertical outlet box. Order separately.

FIGURE 16-29b Microprocessor-controlled thermostat specifications. (Courtesy of Honeywell, Inc.)

bellows to move. As the temperature falls, the inert pressure decreases, and the bellows move in the opposite direction. As required by the design, switching is controlled by the moving end of the bellows.

Limit Pressure Controls   Limit pressure controls are used to connect or disconnect electrical power from condenser system components when the pressure being monitored is above or below the preset limit value. The pressure switch used may be a high-pressure, low-pressure, dual-pressure, or pressure-differential switch. Figure 16-35 shows examples of pressure switches.

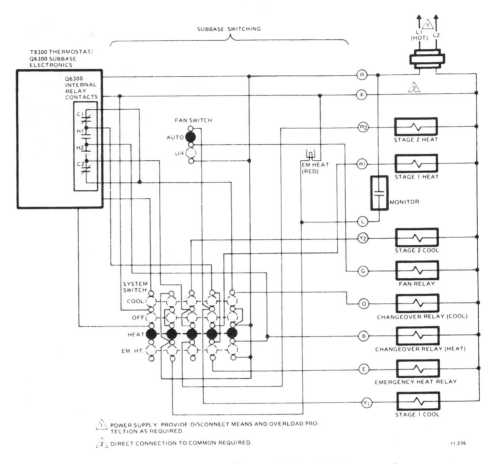

FIGURE 16-30   Wiring connections for T8300B/Q6300B in a two-heat, two-cool heat pump system. (Courtesy of Honeywell, Inc.)

FIGURE 16-31   Heat strip limits.
(Courtesy of BET, Inc.)

FIGURE 16-32   Heat strip. (Courtesy of BET, Inc.)

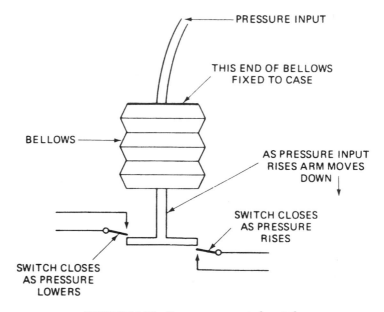

FIGURE 16-33   Pressure-operated switch.

High- and low-pressure switches are similar in appearance. The pressure range at which the switch bellows mechanism operates determines whether it is a high- or low-pressure control. Low-pressure controls in the air-conditioning and refrigeration industry usually have a gauge (psig) range of from 12 inches of mercury (Hg) to 50 pounds per square inch. High-pressure controls usually range from 10 to 425 psig.

Often, the pressure at which the switches actuate is adjustable. There is a difference in the pressure range at which a system should be operational, depending on the altitude at which the system is located or on the desired temperature range.

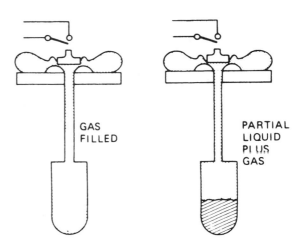

FIGURE 16-34 Temperature-sensitive switches.

FIGURE 16-35 Pressure switches. (Courtesy of BET, Inc.)

OPERATION PROTECTION

Pressure Controls

Pressure controls are used for safety protection with compressor-type condenser systems. A high-pressure control actuates in the system if high-side system pressure exceeds the preset level. When the switch in the pressure control opens, the compressor is stopped. This type of safety control is required to prevent the compressor motor from overloading and to prevent the system from rupture because of excessive internal pressure.

A low-pressure control is used to stop the compressor motor whenever the low-side pressure falls below a preset level. Too low suction pressure most often occurs when there is a refrigerant leak in the system. Damage to the system might result if the unit were allowed to operate under this condition, drawing air and moisture into the system.

If either the high- or low-pressure control opens, the control circuit (24 volts) to the contactor coil is interrupted. The power contacts then open, removing electrical power from the compressor motor.

Figure 16–36 shows the electrical connections of both a high- and a low-pressure control. The switch systems associated with pressure controls are available in many combinations. Pressure controls are available with manual or automatic reset features. Automatic controls reset when the pressure stabilizes and is back within the operating range. If the problem causing the abnormal pressure condition is not corrected, the control again opens.

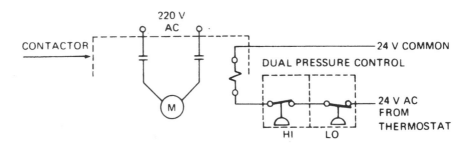

FIGURE 16-36   Dual-pressure control in compressor circuit.

Water-Tower Controls

In special applications, pressure controls may be referred to on schematic diagrams as *reverse acting*. The terminology does not refer to the operation of the pressure control but to the circuitry associated with the control.

For example, consider a reverse-acting fan pressure control in a water-cooled air-conditioning system. In a water-cooled system, a water tower is often used to supply cooled water to the condenser. For the system to operate efficiently, the head (high-side) pressure of the compressor must be maintained within certain limits. If the head pressure becomes too low, the evaporator does not function properly. Low head pressure could be the result of too cool water being supplied to the condenser. To remedy this problem, a reverse-acting fan pressure control connected to the high side of the compressor is used. When pressure is below a set limit, the switch associated with the control opens. This action removes power from the tower fan motor, which is used to cool the water.

The temperature of the water in the tower rises because of heat transfer from the water-cooled condenser. Accordingly, the head pressure rises to the efficient operating level. The pressure switch is again actuated (closed), turning the water tower fan motor back on. This cycle repeats as required; thus, the head pressure is maintained within ideal operational limits.

The reverse-acting fan pressure control operates in the same manner as other pressure controls. Only the name is different; the name refers to the use of the pressure control.

Water Make-Up Control

The water level in a water tower must be maintained at a selected level for best operation of the system. Since water evaporates and is lost to drift, some method must be incorporated in the tower to make up the needed water. In many cases, this is simply a mechanical float system; in others, it is an electrically controlled device.

Make-Up Valve If city (mains) water is used to maintain the level in the tower sump, a common control is the make-up valve. An example of an electric make-up valve was shown in Figure 16-16. Sensing of the water level may be accomplished with a float system or an electric-circuit water-level sensor.

1. *Float system:* The float system consists of a ball-float attached to a movable lever control arm. The float-lever-arm assembly is used to actuate an electric switch. The switch, in turn, controls the application of electric power to the solenoid of the make-up valve. Figure 16–37 is a sketch of a float system.

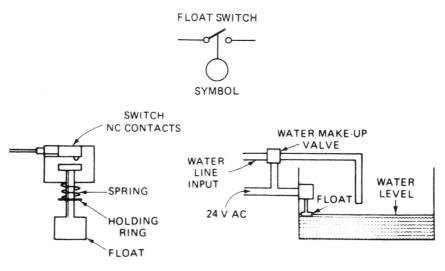

FIGURE 16-37 Float system.

2. *Sensor system:* The electric-circuit sensor system for water level consists of two electrical terminals (Figure 16-38). When there is water (low resistance) between the two terminals, a complete electric path is provided. If the water level falls below the terminals, the electric circuit is interrupted and the high resistance (air) is sensed in the amplifier. A control signal is then sent to the water-valve solenoid, turning the water on.

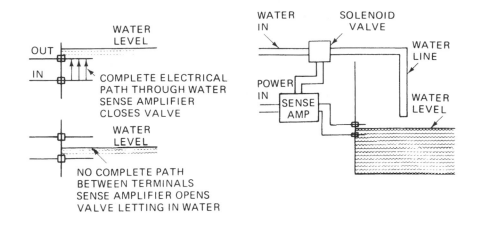

FIGURE 16-38   Using water as a conductor.

SUMMARY

The different types of electrical and mechanical controls available for service in air conditioning and refrigeration are very extensive. It is impractical to attempt to cover them all in a basic textbook. It is therefore recommended that the catalogs and specification sheets of control manufacturers be studied in detail for further information; manufacturers' catalogs and "spec" sheets are an invaluable source of technical information. Figure 16–39 shows a specification sheet selected at random. The application and operational characteristics of the control are clearly stated. This spec sheet is typical of the hundreds available.

PRACTICAL EXPERIENCE

Procedure
1. Locate examples of the different controls covered in operating systems.
2. Observe how the devices are connected in the systems.
3. Be certain that you understand how the devices that relate to personal protection operate.

Conclusion   To understand air-conditioning and refrigeration systems operation, it is necessary to understand how the controls operate.

 PENN PRODUCTS DIVISION · PENN CONTROLS, INC.

SUB-BASE UNIT
Must be used with Series T51 Rimset* Sensing Unit
(Was Series 888)

These instructions cover installation procedures for thermostat sub-base units only. For thermostat adjustment instructions, etc., see instruction sheet packed with thermostat unit.

APPLICATION

The Rimset thermostat is a low voltage, precision temperature regulating instrument for close control of heating, cooling or combination year 'round air conditioning systems.

LOCATION

Recommended thermostat location is approximately 5 feet above the floor, on an inside wall. Do not mount where operation may be affected by concealed warm or cold water pipes, warm air ducts, registers, lamps, fireplaces, drafts from stairways, hallways, etc.

INSTALLATION AND WIRING

All wiring must comply with applicable electrical codes and regulations. To install sub-base proceed as follows: Note: *Make sure sub-base has the correct adjustable heat anticipator for the heating thermostat circuit.*

1. Mount sub-base to wall by means of screws supplied. Before mounting sub-base the excess wire should be pushed back into hole in the wall and hole should be plugged to prevent draft affecting operation of thermostat.

2. Complete wiring to terminals on the baseplate in accordance with the wiring diagram furnished with the equipment. Be sure wires are laid flat, see Fig. 3.

Fig. 2 — Schematic view of thermostat installation.

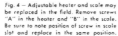

Fig. 3 — Wiring to terminals should be neat and should lie flat so thermostat unit can be "plugged-on" without interference.

Fig. 4 — Adjustable heater and scale may be replaced in the field. Remove screws "A" in the heater and "B" in the scale. Be sure to note position of screw in scale slot and replace in the same position.

* Trademark

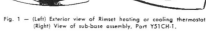

Fig. 1 — (Left) Exterior view of Rimset heating or cooling thermostat. (Right) View of sub-base assembly, Part Y51CH-1.

3. Adjust anticipating heater to correct current setting as indicated by the thermostat current of the valve or relay being controlled, see Fig. 4.

4. "Push-on" the thermostat unit, making sure thermostat unit fits securely on the sub-base.

OPERATION

Selector switches in sub-bases are readily accessible for instant selection of functions. For correct selection of functions, see steps below and schematic wiring diagrams. Refer to diagrams supplied with equipment manufacturer's units.

Heating On

1. Set System selector switch (if provided) to "Heat" position. Heating thermostats without selectors automatically control heating unit.

2. Set Fan selector switch to "Auto" position (Fan selector not provided on sub-bases for heating only).

Cooling On

1. Set System selector switch to "Cool" position or "Auto" position, whichever is indicated on baseplate. Cooling thermostats without selectors automatically control cooling unit.

2. A choice of intermittent or continuous fan operation is available on sub-bases with Fan selector switches:
   a. For fan operation to cycle with the cooling unit, set Fan selector to "Auto" position.
   b. For continuous fan operation set Fan selector to "On" position.

Ventilation without Heating or Cooling

1. Thermostats having system selector switch with "Off" position should be set to "Off."

2. Set Fan selector switch to "On."

FIGURE 16-39 Y51 specifications. (Courtesy of Penn Controls Inc.)

REVIEW QUESTIONS

1. Edison-base fuses are replacing S-type fuses in industry.   T_____   F_____

2. Thermal circuit breakers may be reset immediately.   T_____   F_____

3. Current relays are connected in parallel with motor windings.   T_____   F_____

4. Thermal time relays operate on dc power only.   T_____   F_____

5. Magnetic starters are used to control large motors.   T_____   F_____

6. Magnetic starters with push-button controls have thermal overload contacts in series with the stop-push button.   T_____   F_____

7. Some thermal snap overloads contain an electric heater.   T_____   F_____

8. Thermal snap overloads open if the motor-housing temperature rises above the safe limit.   T_____   F_____

9. Solenoid-controlled valves may be used in water lines to control water flow.   T_____   F_____

10. Connection of input and output lines to a solenoid valve may be interchanged.   T_____   F_____

11. Open thermostats are less prone to trouble than mercury switch types.   T_____   F_____

12. In a temperature-sensitive bellows system, as temperature decreases, pressure increases.   T_____   F_____

13. Anticipators are used in thermostats to decrease the frequency of compressor operation.   T_____   F_____

14. Low-pressure controls interrupt power to the compressor if refrigerant leaks from the system.   T_____   F_____

15. A reverse-acting fan pressure control turns the water tower fan motor off when compressor head pressure reaches a low limit.   T_____   F_____

Semiconductor Devices

OBJECTIVES

Upon completion and review of this unit, you will understand
- What semiconductor materials are.
- The effect of the addition of impurities to making useable semiconductors.
- The action inside a semiconductor diode.
- The electrical characteristics of a diode.
- What zener voltage regulator diodes are.
- How transistors are constructed, and how they operate.
- Special semiconductor devices.

An increase is occurring in the use of semiconductor devices in the control of air-conditioning and refrigeration systems. An example of such a device, the anti-short-cycle timer, is shown in Figure 17-1. The purpose of the device is to keep power from being immediately reapplied to the compressor motor if power is interrupted for any reason. The power is kept from the compressor motor until such time as high head pressure is allowed to bleed down through the evaporator. The device contains transistors, diodes, resistors, capacitors, and a relay. In most cases, the repair of this type of device is not accomplished in the field.

Service technicians should be familiar with the electronic devices associated with air-conditioning and refrigeration control devices to be able to determine whether the device is functioning properly.

DEFINITION

A *semiconductor* is a material that falls somewhere between the good conductors (most metals) and the poor conductors or insulators (such as glass or wax). Semiconductors function in electronic circuits because of their crystalline structure and because they are doped with materials called *impurities*. The impurities provide characteristics to the semiconductor material that make the material useful in rectifiers and amplifiers.

The basic semiconductor materials are germanium and silicon. Silicon is the most popular and is covered here. When the pure silicon material is doped with an impurity, it becomes either *P*-type silicon or *N*-type silicon, depending on the material added as the impurity. The *P* stands for positive and the *N* for negative. The impurity combines with the silicon crystals and modifies the crystalline structure. If *P*-type impurity is added, the crystals of silicon form improperly, leaving a space for an electron in the structure; this space is called a *hole*, and it is movable in the silicon. If *N*-type impurity is added, the crystals of silicon form with an extra electron available; this extra electron is movable.

FIGURE 17-1   Anti-short-cycle timer. (Courtesy of BET, Inc.)

DIODES

A *diode* is a device that allows electrons to flow through it in only one direction. Diodes are useful in converting alternating current into direct current.

In the construction of a diode, *P*-type silicon and *N*-type silicon are joined together (Figure 17-2). Each is electrically neutral before they are joined together. The excess electron in the *N*-type material and the hole (the place for an electron) in the *P*-type material relate only to the crystalline structure. When the two materials are joined together, some of the excess electrons in the *N* type migrate over and fill the holes in the *P* material (Figure 17-3).

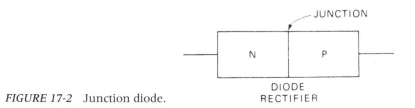

FIGURE 17-2   Junction diode.

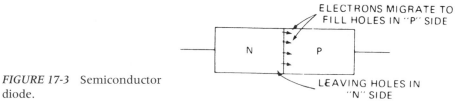

FIGURE 17-3   Semiconductor
diode.

When the electrons from the N side move over to the P side, a potential is developed across the junction. The potential is called a *barrier*. The N material loses some electrons and is positively charged; the P material gains electrons and is negatively charged (Figure 17-4). No further action takes place unless an outside potential is applied to the diode.

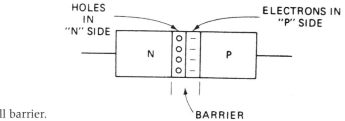

FIGURE 17-4   Potential hill barrier.

If a voltage is applied to the diode, as shown in Figure 17-5, electron current does not flow. Electrons cannot flow because the application of the external potential increases the width of the barrier. The positive terminal of the battery pulls a few more electrons from the N side, increasing the number of holes and widening the barrier. The negative terminal of the battery adds few electrons to the P side, widening the barrier here. The application of the battery potential to the diode increases the resistance of the diode. This is called *reverse bias*.

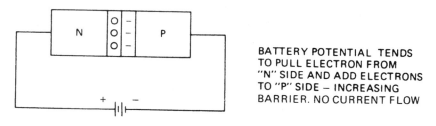

BATTERY POTENTIAL TENDS
TO PULL ELECTRON FROM
"N" SIDE AND ADD ELECTRONS
TO "P" SIDE – INCREASING
BARRIER. NO CURRENT FLOW

FIGURE 17-5   Reverse bias.

If the battery connection is changed, a completely different situation exists (Figure 17-6). The application of the positive to the P material allows for the movement of an electron from the barrier area toward the positive potential. An electron from the N side may move across the junction to fill the hole left by the moved electron in the P side. An electron may travel through the battery and out the negative terminal to the N side of the diode to replace the electron that moved across the junction. The is called *forward bias*.

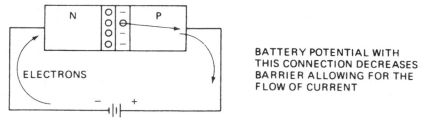

FIGURE 17-6   Forward bias.

The procedure is continuous, and electron current flows through the diode (Figure 17-7). The resistance of the diode is reduced to a low level. The symbol for a diode is shown in Figure 17-7. Electrons flow against the arrowhead.

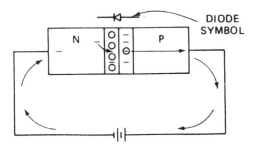

FIGURE 17-7   Conducting diodes.

If an alternating voltage is applied to the diode, the diode conducts every other half-cycle. In the circuit of Figure 17-8, current flows through the diode and R_1 every other half-cycle. The input voltage to the resistor diode combination is represented as E_{in}; the output voltage across R_1 is represented as E_{out}. The circuit is called a half-wave rectifier. A capacitor may be added to the circuit to provide filtering. In Figure 17-9, the connection of the capacitor and the effects on the output waveform are shown.

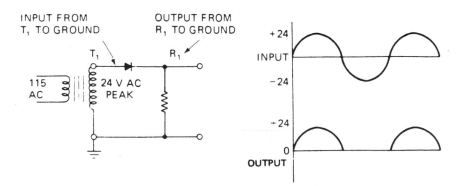

FIGURE 17-8   Half-wave rectifier.

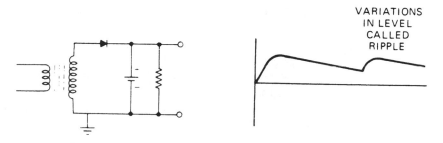

FIGURE 17-9   Filter capacitor maintains voltage.

FULL-WAVE BRIDGES

The full-wave bridge circuit is used to provide an output voltage with less ripple. The circuit is shown in Figure 17-10. Figure 17-10a shows the path for electron flow during the first half-cycle. Figure 17-10b shows the path for electron flow during the second half-cycle. Note that the ripple amplitude has decreased in the filtered output.

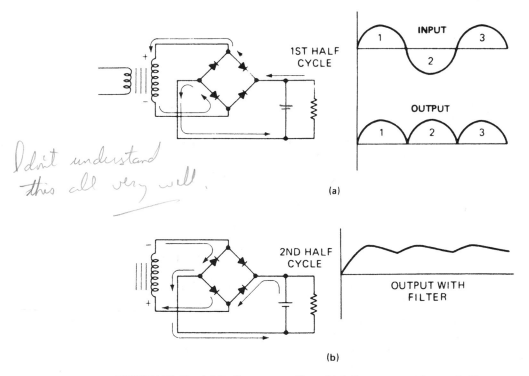

FIGURE 17-10   (a) Full-wave rectifier; (b) full-wave rectifier with filter.

BREAKDOWN VOLTAGE

When a diode is conducting in a circuit, a low-level voltage across the terminals of the diode is present. This is called the diode's forward voltage. If the voltage across the diode is of reverse polarity, the diode will not conduct. There is a maximum reverse voltage that the diode can withstand. If this voltage is exceeded, the diode will break down and conduction will occur. The diode may or may not be damaged, depending on the amount of current flow.

ZENER DIODES

The breakdown voltage across a diode remains fairly constant and is set by the doping ratios. This characteristic provides a means of developing voltage regulators. Voltage-regulating diodes, or *zener diodes*, are diodes developed to operate in circuits where the applied voltage exceeds the diode's breakdown voltage. The diode operates in the breakdown region and has a relatively constant voltage across it. The symbol for a zener diode is shown in Figure 17-11.

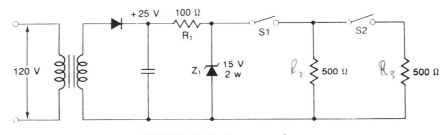

FIGURE 17-11   Zener regulator.

Consider the circuit in Figure 17-11 where a 25-volt dc power supply is connected through R_1, a 100-ohm resistor, to Z_1, a 15-volt, 2-watt zener diode. The switches S_1 and S_2 control the connection of load resistors R_1, 500 ohms, and R_2, 500 ohms. With switches S_1 and S_2 open, the $+25$ volts from the supply exceeds the breakdown voltage of the zener diode. Current will flow up through the zener and the resistor R_1 to the source voltage. The current will rise to 0.1 (100 ma) ampere where 10 volts will appear across the resistor R_1, and 15 volts will be developed across the zener diode. The zener diode has 15 volts across it and 100 ma through it. By Ohm's Law, its resistance is 150 ohms.

When switch S_1 is closed, current will flow through resistor R_2. This would tend to increase the voltage across R_1 and thereby decrease the voltage across the zener. When the voltage across the zener decreases, the zener resistance increases. The current through the zener decreases from 100 ma to 70 ma. With 70 ma through the zener diode and 30 ma through the load resistor R_2, the current through R_1 remains at 100 ma. The 100 ma through R_1 continues to provide a 10-volt drop across R_1, and the output voltage remains 15 volts.

Huh.!?→

[Handwritten margin note top: So how did this voltage "decrease" when R₂ was switched in? This explanation isn't accurate.]

Closing switch S₂ increases the load current by another 30 ma to a total of 60 ma. The current through the zener drops to 40 ma. The current through R₁ remains at 100 ma, and the output voltage remains at 15 volts.

[Handwritten margin note: Note:]

Actually, the voltage across the zener will decrease slightly as the current drawn by the load is increased. It is the slight decrease in voltage across the zener that provides for the change in zener resistance.

[Handwritten: → Confirm this.]

TRANSISTORS

Transistors are three-terminal semiconductor devices that can be used to amplify signals. Two-types of transistors are available, the *PNP* transistor and the *NPN* transistor (Figure 17-12). The *N* and *P* refer to the type of impurity used in the silicon.

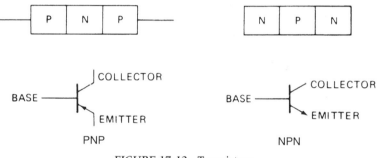

FIGURE 17-12 Transistors.

The main characteristic of transistors is that a small current flow in the emitter-base circuit controls a larger current flow in the emitter-collector circuit. Transistors, like diodes, require the proper polarity of applied voltage for proper operation. The emitter-base junction must have voltage applied in the forward direction for conduction. The applied voltage must oppose conduction in the base-collector junction. An example of transistor amplifiers is shown in Figure 17-13. Note the reverse polarity of applied voltages on the *NPN* and *PNP* circuits.

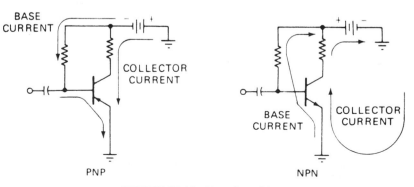

FIGURE 17-13 Transistor bias.

In Figure 17-14, the 100k resistor provides for a small emitter-base current. A larger emitter-collector current also flows. An input signal varies the emitter-base current, and an amplified version of the current flows in the emitter-collector circuit. The varying collector current produces a varying voltage across the 1k collector resistor.

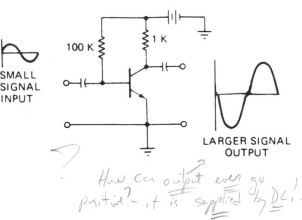

FIGURE 17-14 Transistor amplifier.

How can output ever go positive? - it is supplied by DC!!

INTEGRATED CIRCUITS

Miniaturization processes have provided for the manufacture of complete amplifier circuits in a single, small package. Examples of amplifiers in the form of integrated circuits are shown in Figure 17-15. There are more components in the integrated circuit than in the anti-short-cycle unit that was shown in Figure 17-1.

FIGURE 17-15 Integrated circuit.
(Courtesy of BET, Inc.)

SILICON-CONTROL RECTIFIERS AND TRIACS

A silicon-control rectifier (SCR) is a three-terminal device that has two states of operation, off and on. The symbol of a silicon-control rectifier, a circuit example, and the voltage waveforms of input, the gate, and the output are shown in Figure 17-16. The signal that allows conduction is a positive reference between the gate and cathodes. When there is a positive reference on the gate, the SCR conducts whenever the anode is positive. Note that the positive half-cycle of the input appears in the output only during the positive gate period.

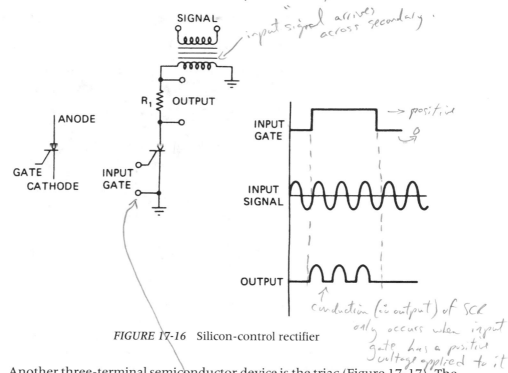

FIGURE 17-16   Silicon-control rectifier

Another three-terminal semiconductor device is the triac (Figure 17-17). The triac allows for conduction in both directions when the gate signal is positive with reference to input main terminal 1.

The triac circuit shown in Figure 17-16 is used in many modern control devices. One such device is the dual time delay shown in Figure 17-18. The dual time delay provides time delay in turning a circuit on and time delay in turning the circuit off. Time delay in control of the inside blower motor is useful in air-conditioning and heating systems.

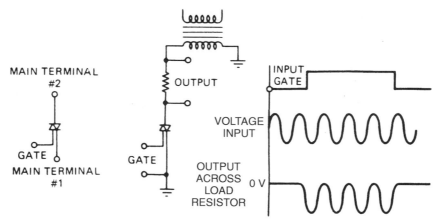

FIGURE 17-17   Triac, controlled output.

FIGURE 17-18   Dual time-delay relay. (Courtesy of BET, Inc.)

As a practical situation, it is proper to delay turning on the inside blower motor until after a heating system has had time to heat the plenum area of the furnace. Otherwise, cold air in the plenum would be moved into the room to be heated. It is also proper to continue to move the heat in the plenum into the space to be heated after the heating device has been turned off.

The circuit in Figure 17-19 shows how this might be accomplished in a gas furnace system. The dual time delay is set for 15-second delay on and 25-second

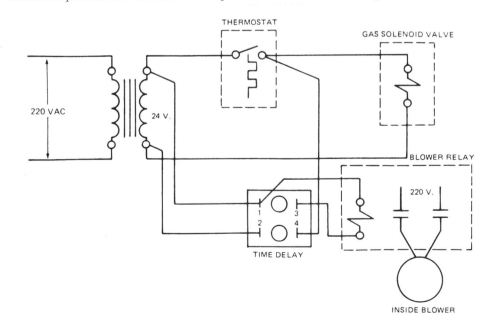

FIGURE 17-19   Connection of time delay.

delay off. (The required delay depends on the particular system.) When the thermostat calls for heat, the gas solenoid valve is actuated, turning the heater on. Approximately 15 seconds later, after the plenum has had time to heat, the time-delay relay energizes the inside blower relay. The contacts of the inside blower relay feed power to the blower motor. The operating blower moves warm air from the plenum to the space to be heated.

When the room reaches the selected temperature, the thermostat opens. The gas solenoid valve de-energizes, turning the furnace heating unit off. The time delay keeps the blower relay energized. The blower continues to move the heated air from the plenum to the area to be heated. The temperature of the plenum reduces. After 25 seconds, the time delay causes the blower relay to de-energize, and the inside blower motor stops. A considerable amount of energy may be saved using a time-delay system.

TEMPERATURE SENSING

The materials that are used to make up resistors are usually temperature sensitive. Most materials increase in resistance as temperature increases. There are, though, special resistors that are designed to have a relatively large variation in resistance as temperature changes. They are called temperature-sensitive resistors (see Appendix A.)

Temperature-Sensing Element

A special temperature-sensitive semiconductor device has recently been developed. The device produces a linear variation in output voltage with temperature. Connected as in the circuit of Figure 17-20, a linear output voltage of $+10$ mV/°F may be obtained for temperatures between $+5$ to $+300$ °F.

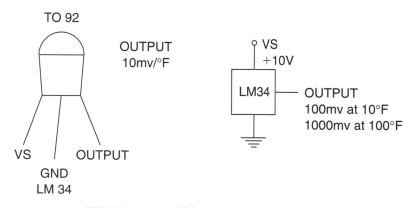

FIGURE 17-20 Solid-state temperature sensor.

COMPRESSOR-MOTOR PROTECTORS

Compressor-motors are designed to operate within a fixed temperature range. They are also designed to operate within a fixed current range. (See Figures 16-10 through 16-12 for examples of thermal overloads and their connection.)

Under special conditions, the thermal overload can become an active element in compressor failure. This is not to suggest that the overload should be removed. The compressor could burn out very rapidly without the overload. It is just that the overload is active in the operation.

described
immediately
following

Low Voltage (Brownout)

Low-voltage effect on compressors is dependent on how close to its rated load the compressor is operating.

Underloaded Compressor

Consider an underloaded compressor. This compressor is drawing less than full-load current during normal operation. As line voltage lowers the current drawn by the compressor motor will decrease. The decreasing current flow with lowering voltage will continue until a point is reached where required power is greater than available power. The current will then increase rapidly with lowering voltage.

Fully Loaded Compressor

With a fully loaded compressor, current flow will immediately start to increase with a lowering of line voltage.

Effect of Increasing Current During Brownout   Increasing current flow in the compressor causes increased heat. The increased heat, along with the increased current flow through the thermal overload, causes the overload to open.

After a period of time, the overload cools and the contacts close. The compressor tries to start. If the voltage is too low, the compressor cannot start. High current is drawn and the overload again opens. This process may continue as

1. overload opens;
2. overload cools;
3. overload closes;
4. compressor tries to start;
5. high current is drawn;
6. overload opens;
7. repeat 2 to 6.

If this process is allowed to continue, the compressor will finally burn out.

Low-Voltage–Anti-Short-Cycle Cutout

Recent developments include the low-voltage–anti-short-cycle cutout. The device is designed to provide a delay in the application of power to the compressor after it becomes available to the system. The device also removes power from the compressor and keeps it from the compressor if voltage supplied is 10% below the specified voltage. Most compressor motor specifications allow for 10% voltage decrease where minimum operating voltage is given.

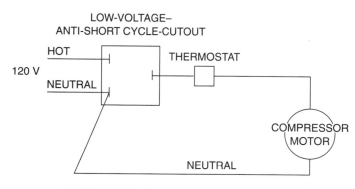

FIGURE 17-21 Compressor motor protector.

The circuit for connection of the low-voltage–anti-short-cycle device is given in Figure 17-21. A photograph of a low-voltage–anti-short-cycle cutout is given in Figure 17-22.

FIGURE 17-22 Low-voltage–anti-short-cycle cutout. (Courtesy of BET, Inc.)

There is a delay of approximately 4 minutes between the application of power to the device and power to the compressor.

If at any time the line voltage decreases below 105 V, power will be removed from the compressor. Power will not be returned to the compressor until 4 minutes after voltage above 105 V is supplied to the system.

LIGHT-EMITTING DIODES (LEDs)

All semiconductor diodes produce light (photons) when conducting. The light is not seen because the material silicon (Si) or germanium (Ge) is opaque. The photons are produced but do not escape the device.

Light-emitting diodes (LEDs) are constructed of materials that are translucent to light. The light can escape and therefore can be seen.

One of the earliest LEDs available was in the color red. Shortly after the red LED came green LEDs. LEDs are now available in white, red, orange, yellow, green, and the latest addition, blue. Light-emitting diodes (LEDs) are used as indicating devices. They are often used to indicate the on-off condition. They are small, bright, and produce little heat. See Figure 17-23.

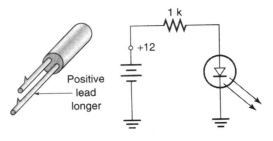

FIGURE 17-23 Light-emitting diode.

PHOTOCONDUCTIVE CELLS

Photoconductive cells are usually made using cadmium sulfide (CdS) or cadmium selenium (CdSe) as the cell's active material. The light-sensitive material is deposited on a glass surface. Two electrodes are fixed over the glass substrate and provide a controlled path for light to act on the active surface.

The resistance of a photo cell is high in darkness and low when light strikes its surface. The symbols used for photoconductive cells are shown in Figure 17-24. The arrows in Figure 17-25 indicate the light that the device is sensitive to. The second symbol used the Greek letter *lambda* (λ) to represent the wavelength of light. See Appendix A, Schematic Symbols.

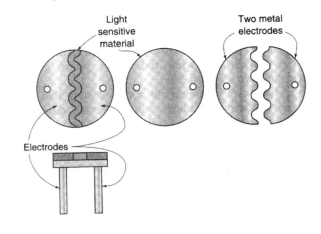

FIGURE 17-24 Photo-conductive cell.

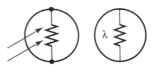

FIGURE 17-25 Photoconductive cell symbols.

PHOTOVOLTAIC CELLS

The photovoltaic cell (Figure 17-26) converts light energy directly into electrical energy. It is a junction device made from a P-type layer and an N-type layer. When light strikes one of the junction surfaces while the other is isolated from the light, a voltage is produced across the junction.

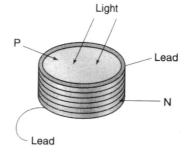

FIGURE 17-26 Photovoltaic cell.

PHOTOTRANSISTOR

By incorporating a photovoltaic cell in the same body as a transistor a light-sensitive transistor may be produced. The equivalent circuit is shown in Figure 17-27.

When light strikes the surface of the photodiode within the transistor, it produces a voltage which turns the transistor on. The greater the amount of light striking the phototransistor, the greater the collector current.

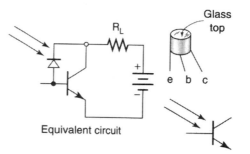

FIGURE 17-27 Phototransistor.

SUMMARY

There are many uses for semiconductors in control devices. The popularity of semiconductors is increasing. The air-conditioning and refrigeration technician should be familiar with these devices and be able to determine whether they are operating properly or are malfunctioning.

Whenever a control does not perform the required function, it should be checked for proper inputs. These inputs usually are a supply voltage and the control signals. If these two are present but the semiconductor device does not perform according to specifications, it should be replaced. Repair is seldom accomplished in the field.

PRACTICAL EXPERIENCE

Required equipment
1. Analog volt-ohm-milliammeter (VOM)
2. Transistor, NPN, such as 2N3904

Procedure
1. Select resistance on the multimeter and select the highest R× scale, probably R × 10,000.
2. Connect the black lead (negative) to the transistor emitter.
3. Connect the red lead (positive) to the transistor collector.
4. What does the meter read? _____(See Note 1.)
5. Wet one of your fingertips.
6. Place your fingertip between the transistor collector and the base lead, Figure 17-28. (See Note 2.)
7. What does the meter read? _____(See Note 2.)
8. In step 3 the transistor is turned (off/on).
9. In step 6 the transistor is turned (off/on).
10. Why, in your opinion, is there a difference (if any) in steps 3 and 6? (See Note 3.)

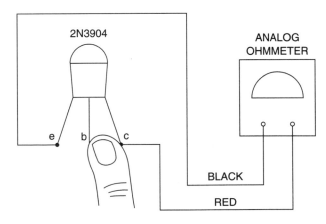

FIGURE 17-28 Finger used as collector, to base resistor.

Notes

1. In Step 3 the meter should indicate (high resistance), full scale, or close to full scale, since there is no emitter base current to turn the transistor on.

2. In Step 6 the meter should jump up the scale, indicating a lower resistance. The transistor is turned *on*.

Explanation   When the wet fingertip is placed between the collector and base, current flows through the skin; from the positive (+) meter lead through the transistor base. When emitter-base current flows, the transistor turns *on*, lowering its resistance. The battery in the analog meter provides the necessary power to operate the transistor.

Conclusions

- Semiconductors have resistance that falls between good conductors and good insulators.
- "Doped" semiconductor materials are useful in the manufacture of diodes and transistors.
- Diodes have low resistance in one direction and high resistance in the other direction.
- In a transistor a small base current controls a larger collector current.

REVIEW QUESTIONS

1. The two types of material used in semiconductors are cobalt and silicon.   T_____   F___✓___

2. Diodes are used to change dc to ac.   T_____   F___✓___

3. Diodes normally conduct in both directions.   T_____   F___✓___

4. The output of a full-wave rectifier is easier to filter than the output of a half-wave rectifier.   T___✓___   F_____   *produces a more regular output.*

5. Transistors are two-terminal semiconductors.   T_____   F_____

6. In a transistor, a small signal current in the collector controls a large signal current in the base.   T_____   F___✓___

7. Transistors may be used to make amplifiers.   T___✓___   F_____

8. Complete amplifiers may be formed as miniature integrated circuits.   T___✓___   F_____

9. Silicon-control rectifiers allow for current flow in both directions.   T_____   F___✓___

10. Triacs require a negative gate for turn-on.   T_____   F___✓___

Unit 18

Air-Conditioning Circuits

OBJECTIVE

Upon completion and review of this unit, you should be familiar with air-conditioning circuits, including the components that make up the circuits.

A number of combinations of components and circuits are used in the air-conditioning and refrigeration industry. The service technician should become familiar with these circuits and the components most often used. One of the most reliable sources for information on various components is the component manufacturers' catalogs and specification sheets. The study of electrical systems often includes references to equipment manufacturers' service bulletins.

In this unit, basic air-conditioning system control will be covered in detail, since the greater number of controls is associated with air-conditioning systems. The electrical control of refrigeration systems is similar to air-conditioning and is usually more simple and direct. See Appendix A for symbols common to air-conditioning systems.

UNIT (COOL): 7000 BTU/HOUR

The electrical circuits of an air-conditioning system are illustrated as an aid to the service technician, as a schematic in Figure 18-1 and as a ladder diagram in Figure 18-2. Although these illustrations appear to be different, they present the same electrical system. The tie points and wire connections have been coded in each illustration to show the actual similarities. Since both types of illustrations are found in the field, examples of each are covered in this unit.

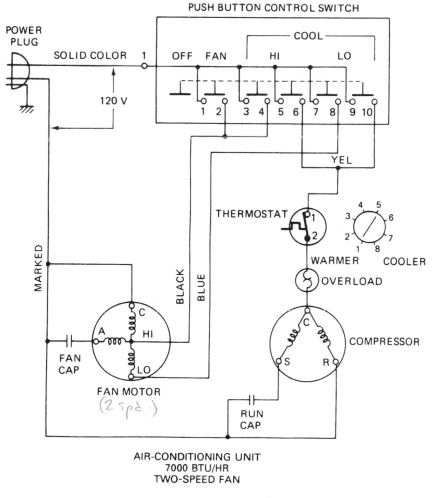

FIGURE 18-1 Unit off.

The push-button control switch provides a means of selecting OFF, FAN, HI-COOL, or LO-COOL. When the OFF button is pressed, all other buttons are released and power is removed from all components of the air-conditioning unit.

Fan

When the FAN button is depressed, power is fed from the wall plug to switch terminal 1. Terminal 1 is mated to terminal 2, as shown in Figure 18-3. Power is fed through the switch and on the solid lead to the HI input to the fan motor. The (C) common terminal of the fan motor is connected to the neutral side of the 120-volt input power source.

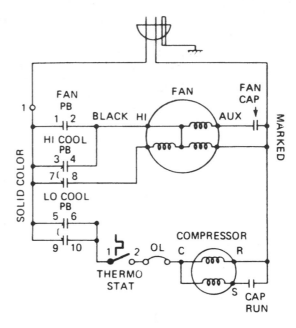

FIGURE 18-2 Ladder diagram.

Electric power is also fed from the HI input of the fan motor through the auxiliary winding (A) to the fan capacitor. The other side of the fan capacitor is connected to the marked input of the power source.

The fan motor operates at high speed whenever the FAN button is depressed.

Hi-Cool

When the HI-COOL button is depressed (Figure 18-4), electric power is fed from the hot line through contacts 3 and 4 of the HI-COOL switch to the HI input of the fan motor. The fan motor operates just as it did with the FAN button depressed.

In the second section of the HI-COOL switch, contacts 5 and 6 provide a path for power from the input line to terminal 1 of the thermostat. When the return air from the room is above the selected temperature, contacts 1 and 2 of the thermostat are mated. Thermostats commonly used with room air conditioners are controlled by sensing the temperature of the air returned from the room. The thermostat temperature sensor is at the inlet side of the evaporator.

Electric power is fed through the thermostat and the thermal overload to the common (C) terminal of the compressor motor. The run (R) winding is connected back to the marked input power line. The start (S) winding of the compressor is returned to the marked line through the run capacitor.

The compressor motor operates, providing the necessary pumping action. The fan motor is operating at high speed, providing the maximum air movement through the condenser, as well as maximum air movement over the evaporator. Thus, cool air is provided to the room. The fan motor is usually of the two-shaft type with the evaporator fan blade (or blower wheel) and the condenser fan blades at opposite ends.

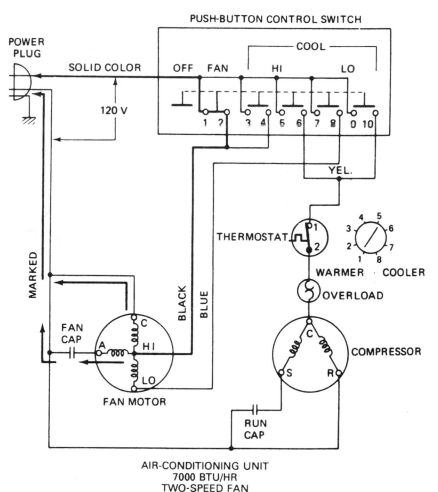

FIGURE 18-3 FAN selected.

Lo-Cool

When the LO-COOL button is depressed (Figure 18-5), electric power to the fan motor is fed from the hot line through contacts 7 and 8 of the LO-COOL switch to the LO input of the fan motor. With the LO winding connected in series with the HI winding, motor speed is decreased. The return paths to the marked line are the same for HI-COOL operation. A low volume of air moves across the condenser and the evaporator sections of the system with the fan motor operating at low speed.

The circuit to the compressor motor from the input line is through contacts 9 and 10 of the LO-COOL switch and contacts 1 and 2 of the thermostat. The compressor operates in the same manner as it did in HI-COOL.

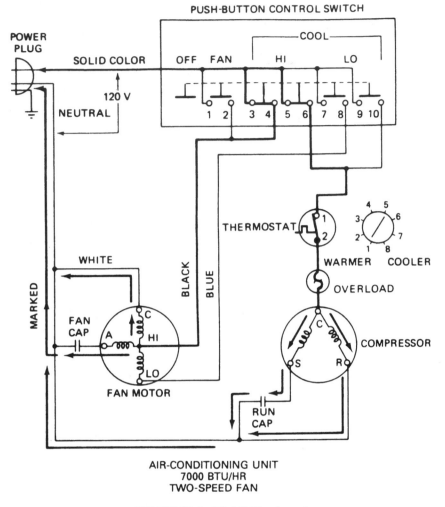

FIGURE 18-4 HI-COOL selected.

The thermostat may be adjusted to a selected "comfort" temperature. The compressor operates only if the room temperature is above the selected temperature.

LARGE WINDOW UNIT: COOL

A larger—18,000-Btu (British thermal unit)—air-conditioning unit with three fan speeds is shown in Figure 18-6. The unit includes push-button control of fan alone at high speed and cool with high, medium, or low fan speeds. Because of starting torque requirements, the compressor employs a start capacitor. The removal of the start capacitor after the compressor reaches a selected speed is accomplished by use of a potential relay.

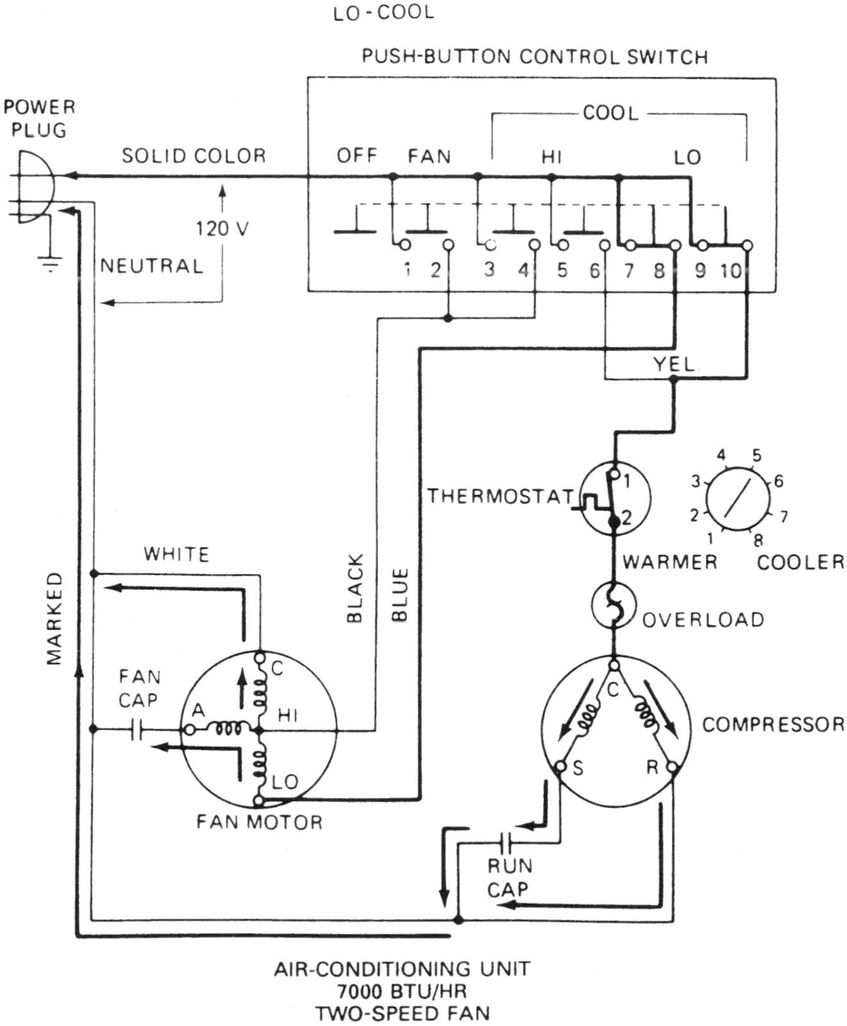

FIGURE 18-5 LO-COOL selected.

The circuit for fan operation is the same as shown in Figure 18-3. Power
is fed in on the black lead to contact 1 of the switch, through the switch to contact 2, and from there to the HI-speed input terminal of the fan motor. The
common terminal of the fan motor is connected to the other input power lead,
the red lead. The auxiliary winding is also returned to the red lead through the
fan capacitor.

Hi-Cool

When the HI-COOL button is depressed, the fan operates. The compressor starts
and runs if the room air temperature is above that selected at the thermostat.

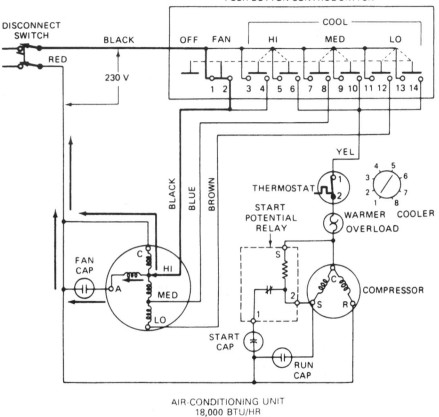

FIGURE 18-6 FAN only, three-speed.

The circuit for the fan and the start circuit for the compressor are shown in Figure 18-7. Electric power to the fan is fed through contacts 3 and 4 of the selector switch. Electric power to the compressor is fed through contacts 5 and 6 of the switch.

If the room air temperature is above the selected temperature, the contacts of the thermostat switch are closed. Power is fed through the thermostat and the bimetal thermal overload to the common terminal of the compressor. The run *(R)* terminal of the compressor is connected to the red input power lead. The start *(S)* terminal of the compressor is connected to the red input power lead with the start and run capacitor in parallel. The circuit for the start capacitor is through the start relay terminal 2, to the closed contacts of the relay to terminal 1, and to the start capacitor.

The compressor motor starts and increases in speed. As the compressor motor reaches correct operating speed, the voltage across the start *(S)* winding is high enough to cause the potential relay to energize. The contacts open, and the circuit (from terminals 2 to 1) of the relay is removed. The start capacitor is electrically removed from the circuit.

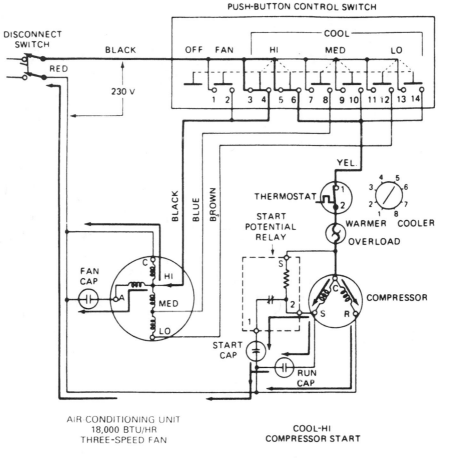

FIGURE 18-7 Compressor starting.

The compressor motor circuit with the compressor at operating speed is shown in Figure 18-8. The start capacitor is not connected in parallel with the run capacitor since the relay contacts are now open. The fan operates at high speed with power though contacts 3 and 4 of the select switch.

Operation of the compressor with the fan in the MED and LO speeds is similar to HI operation. Electric power is fed to the compressor through contacts 9 and 10 of the select switch when MED is chosen or contacts 13 and 14 at the select switch when LO is pressed.

The fan motor operates at medium speed with electric power through contacts 7 and 8 of the select switch or at low speed with electric power through contacts 11 and 12 when LO is selected.

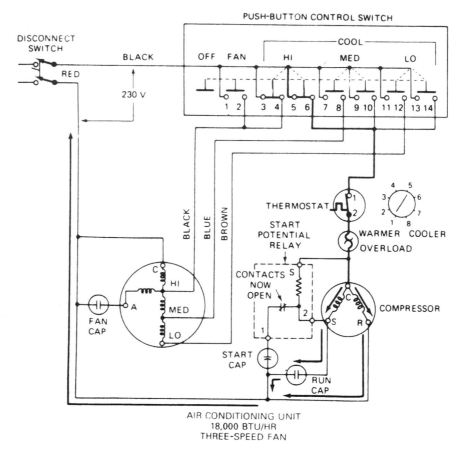

FIGURE 18-8   Compressor running.

THERMOSTAT: SINGLE-STAGE HEAT/COOL

A wiring diagram of a typical thermostat used for remote sensing of room temperature is shown in Figure 18-9. This is a room thermostat that can be used to control a single-stage heat pump with added heat strip. The low-voltage thermostat shown is of the push-button select type.

The required low voltage (24 volts) is supplied by a step-down transformer (Unit 12). One side of the low-voltage supply is connected to terminals R and T (often designated R_c and R_h) of the thermostat. The outputs of the thermostat, Y, W, and G, are connected to the actuating coils of the compressor contractor, heat relay, and fan relay, respectively. The actuating coils of these components are returned to the common terminal of the control transformer.

Thermostat terminal identification usually indicates the color coding of the low-voltage wire used. It also aids in circuit identification, as follows:

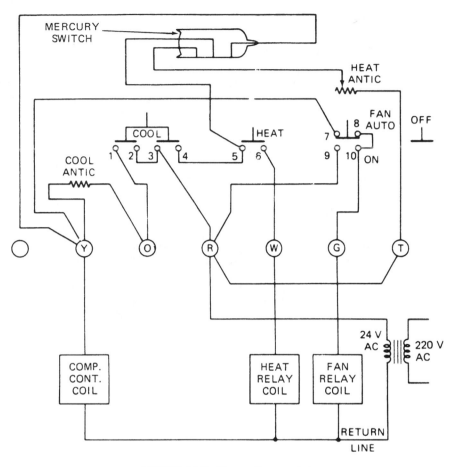

FIGURE 18-9 Room thermostat.

R, red: thermostat input

R_c, red: thermostat input (cooling)

R_h, red: thermostat input (heating)

G, green: fan (blower) motor circuit

Y, yellow: cool circuit

Y_1, yellow: cool circuit, stage 1

Y_2, yellow/TR: cool circuit, stage 2

W, white: heat circuit

W_1, white: heat circuit, stage 1

W_2, white/TR: heat circuit, stage 2

Fan On

When the FAN ON button is depressed, 24-volt ac is connected from the input line to the R input terminal (Figure 18-10). From the R input terminal, the circuit is through the closed contacts 9 and 10 of the FAN ON switch to the G output terminal of the thermostat. In this position, the fan runs at all time.

Heat: Fan Auto

When the HEAT button is depressed and the FAN button is in AUTO, 24-volt ac is connected from the T input through the heat anticipator resistor to the left contact of the mercury bulbs (Figure 18-11). The (T) terminal is the 24-volt ac input to the thermostat for heating. If room temperature is below the selected temperature (heat required), the mercury mates the center contact to the left contact. The circuit is complete through the mercury bulb to contact 5 of the heat switch. The heat switch is closed; the circuit is complete from contact 5 to 6 to the W output

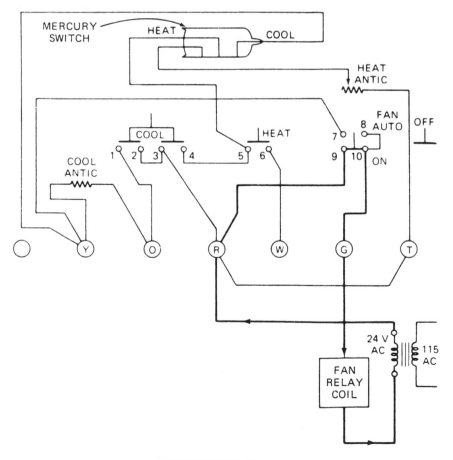

FIGURE 18-10   Fan on.

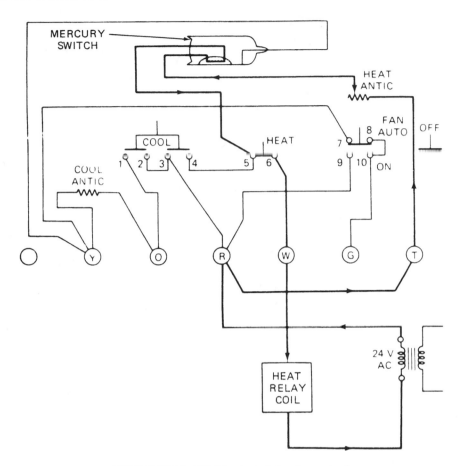

FIGURE 18-11   HEAT selected; heat required.

of the thermostat. The heat relay energizes and provides heat to the system. Control of the inside fan is through the use of a set of contacts on the heat relay, a heat-sensing device in the furnace plenum, or some other source outside the thermostat. Inside fan control systems are covered later in this unit.

When installing a thermostat to control an air-conditioning system providing cooling and heating, the technician should refer to the thermostat manufacturer's catalog for the proper subbase (if required) for that system combination.

Cool: Fan Auto

Cool   When COOL is selected, contacts 1 and 2 and contacts 3 and 4 of the cool switch are mated (Figure 18-12). The 24-volt ac is fed from the transformer to the R input terminal of the thermostat. From the R input terminal, the connection is to contact 3 of the cool switch. Contact 4 is mated to 3, and the circuit is complete

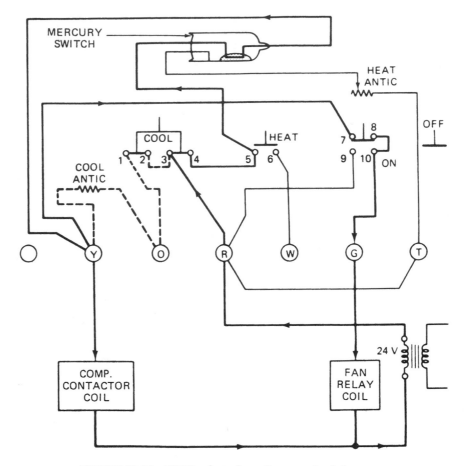

FIGURE 18-12 COOL selected; cooling required; fan auto.

through the switch, where a connection is made to the center terminal of the mercury bulb. If cooling is required, the mercury connects the center terminal to the
left terminal, completing a path to the Y output of the thermostat. The compressor contactor coil is connected from the Y output of the thermostat to the 24-volt
ac return.

Anticipator The circuit of the cool anticipator is from contact 3 to 2 to 1 of the
cool switch to contact 0 of the thermostat to the cool anticipator resistor. When the
circuit through the mercury bulb is open, heat is generated in the resistor. When
the circuit through the mercury bulb is complete, the cool anticipator resistor is
shorted out. The operation of heat and cool anticipators was covered in Unit 17.

Fan Auto With the fan switch in AUTO and the mercury bulb providing for operation of the compressor, the circuit to the fan relay is completed from Y of the
thermostat to contact 7 of the fan switch. Contacts 7 and 8 of the fan switch are
mated, and a circuit is completed to the G output of the thermostat. The fan relay
coil is connected between G of the thermostat and the 24-volt ac return.

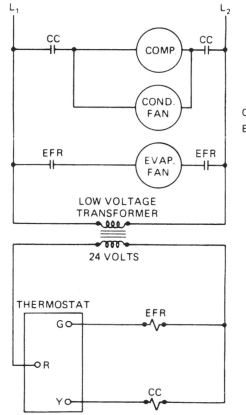

CC – COMPRESSOR CONTACTOR
EFR – EVAPORATOR FAN RELAY

FIGURE 18-13   Air-conditioning, straight COOL.

SIMPLIFIED SCHEMATICS: LADDER DIAGRAMS

Simplified schematics provide a means for understanding systems operations. A simplified schematic of a straight cool system is shown in Figure 18-13. The high- and low-voltage sections of the circuit are separated by the low-voltage transformer. The 24-volt control is fed into the thermostat at the R terminal. Action with the thermostat provides for 24-volt ac output at terminals G and Y, as required. Whenever there is 24-volt ac at G, the evaporator fan relay energizes. Whenever there is a 24-volt ac at Y, the compressor contactor energizes.

On the high-voltage side of the circuit, the compressor, condenser fan, and evaporator fan are shown. When the compressor contactor is energized, both the compressor and the condenser fan operate. When the evaporator fan relay is energized, the evaporator fan operates.

Cool with Reverse-Cycle Heat

The outputs of the thermostat in Figure 18-14 control the evaporator fan relay, the compressor contactor, and the heat relay. A set of normally open contacts of the heat relay bypasses the thermostat in heat operation to provide an energizing circuit to the compressor contactor. The bypass circuit also provides for a 24-volt ac input to the thermostat at the Y terminal. The effect of this connection is an energizing circuit to the fan relay through the AUTO section of the fan switch terminals 7 and 8. The thermostat circuit is shown in Figure 18-15.

The high-voltage circuits in Figure 18-14 are controlled by contacts of the compressor, heat relay, and evaporator fan relay. The heat relay provides for the energizing of the reverse-cycle solenoid.

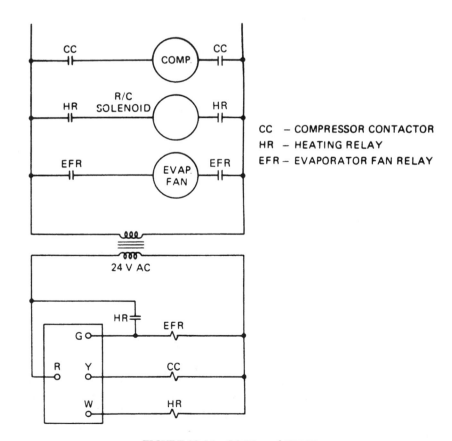

CC — COMPRESSOR CONTACTOR
HR — HEATING RELAY
EFR — EVAPORATOR FAN RELAY

FIGURE 18-14   COOL and HEAT.

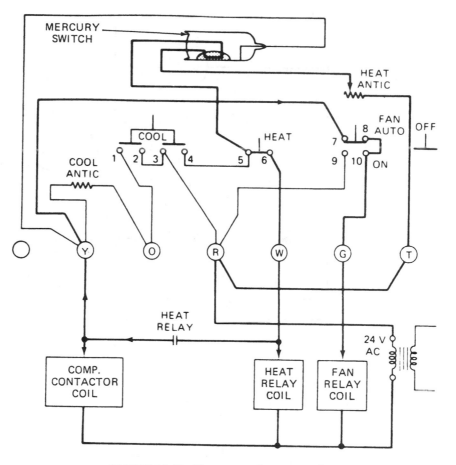

FIGURE 18-15 Heat pump, heat operation.

Water-Cooled Condenser/Strip Heat

A simplified schematic of an air-conditioning system with a recirculating water-cooled condenser and strip heat is shown in Figure 18-16. The low-voltage circuits are changed with the addition of a water pump relay coil connected in parallel with the compressor contactor coil. Whenever the compressor contactor is energized, the water pump relay is energized.

In the high-voltage circuits, the compressor contacts control the compressor. The evaporator fan relay contacts control the evaporator fan. The water pump relay contacts control the water pump. When all the contacts are closed, the system operates.

The heat relay contacts provide a circuit to the strip-heat unit. When the heat relay is energized, a second set of contacts completes a circuit to the evaporator fan motor.

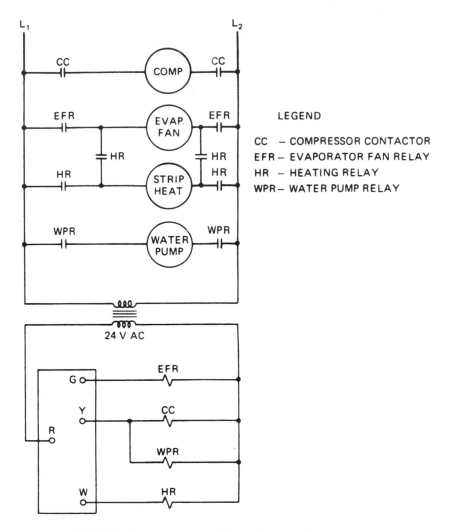

FIGURE 18-16   Water-cooled condenser and strip heat.

Straight Cool/Heat Strip

A schematic of a basic straight cool system with heat strip is shown in Figure 18-17. In low-voltage circuits, the thermostat action is straightforward. The coils of the evaporator fan relay, the compressor contactor, and heat relay are controlled directly by outputs of the control thermostat.

    The only variation of the high-voltage circuit is the addition of a set of contacts of the heat relay in parallel with the evaporator fan relay contacts. The evaporator fan operates whenever either the heat relay or the evaporator fan relay is energized.

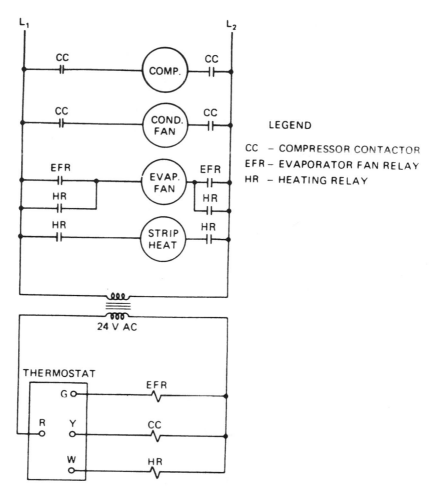

FIGURE 18-17   Straight cool with heat strip.

SIMPLIFIED SYSTEM SCHEMATIC

The circuit shown in Figure 18-18 is also a straight cool system with strip heat. The operation of the evaporator fan relay, heat relay, and compressor contactor is the same as previously described. The additional control contacts shown are in the compressor contactor, low-voltage circuit. They are safety devices used for system protection in equipment operation. The high-pressure control (HP) and the low-pressure control (LP) are designed to open whenever a preset pressure limit is exceeded.

Lock-Out Relay The compressor contactor coil de-energizes whenever any set of contacts in series with the contactor coil opens. The lock-out relay energizes at the same time, and its normally closed contacts open. A completed circuit to the contactor coil is not possible even if the original set of open contacts again closes.

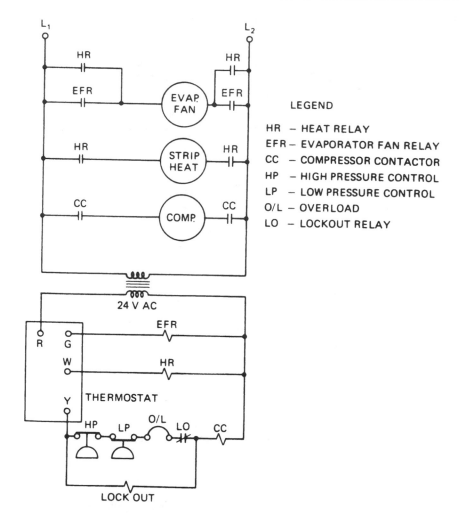

FIGURE 18-18 Straight cool with heat strip and lock-out protection.

The lock-out relay coil has a high impedance, whereas the compressor contactor coil has a low impedance. If the high-pressure control is open, as shown in Figure 18-19, the circuit from the 24-volt transformer runs through the thermostat to the high-pressure control (now open). The circuit continues to the lock-out relay coil through the coil and to the compressor contactor coil. The circuit is completed through the compressor contactor coil to the return side of the 24-volt transformer. Since the lock-out relay coil impedance is high, most of the 24 volts appear across this coil; the lock-out relay energizes. Very little voltage appears across the low-impedance coil of the compressor contactor, causing it to de-energize. The normally closed contacts of the lock-out relay open, ensuring that the compressor contactor does not energize again until the 24 volts supplied to the circuit is removed and the lock-out relay de-energizes.

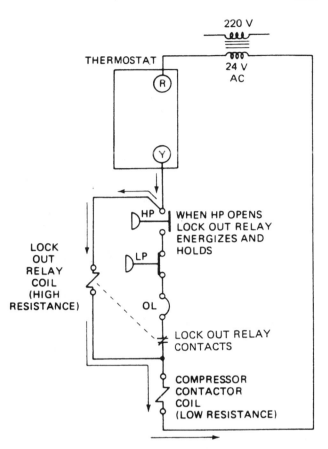

FIGURE 18-19 Lock-out re-
lay control.

Removal of the 24-volt ac source to the lock-out relay may be accomplished by selection at the thermostat or by opening of the main power switch.

Thermostat: Two-Stage Heat, Two-Stage Cool

The diagram in Figure 18-20 is that of a two-stage heat, two-stage cool thermostat. The thermostat provides for switch selection of fan, AUTO or ON, and system, AUTO, HEAT, or COOL.

The diagram is shown with the system OFF and the fan ON. The 24-volt ac input signal is fed from the R input terminal through the fan ON-switch section to the G output of the thermostat.

Cool Selected   In Figure 18-21, the thermostat is shown with the fan in AUTO and the system in COOL. The heavy solid lines indicate the electrical circuit to Y_1 through the thermostat for the 24-volt ac control voltage when single-stage cooling is required.

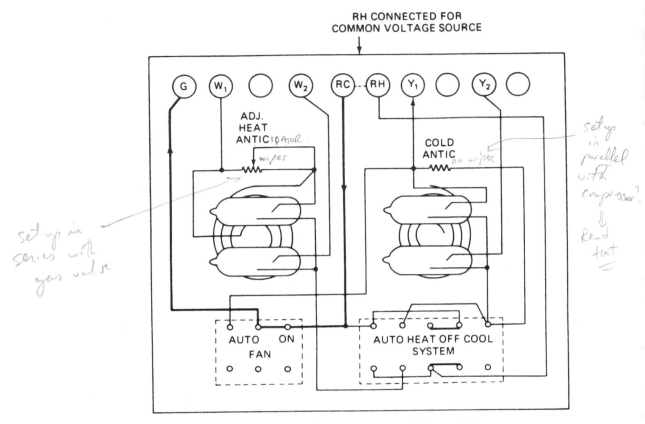

THERMOSTAT—TWO-STAGE HEAT, TWO-STAGE COOL

FIGURE 18-20 Fan on.

There is an approximately 2° F (1.1° C) temperature differential (Td) between first-stage cooling and second-stage cooling. The dashed line shows the electrical circuit to the Y_2 output when two-stage cooling is required.

Heat Selected In Figure 18-22, the thermostat is shown with the fan in AUTO and the system in HEAT. The heavy solid line indicates the electrical path to W_1 through the thermostat for the 24-volt ac control voltage when single-stage heating is required.

There is an approximately 2° F (1.1° C) temperature differential between first-stage heating and second-stage heating. The dashed line shows the electrical circuit to the W_2 output.

The operation of the fan when heat is selected is controlled by a heat relay in the low-voltage circuit that is not a part of the thermostat.

Auto Selected When the system select switch is placed in the AUTO position, the air-conditioning system provides heating or cooling as required to maintain the selected temperature.

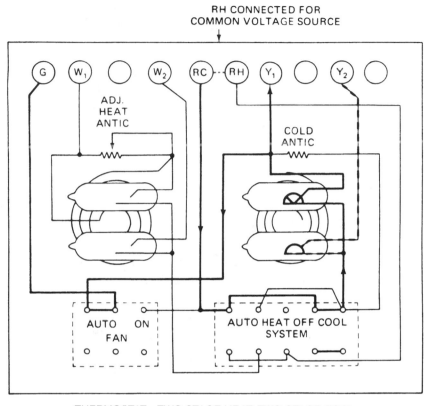

THERMOSTAT—TWO-STAGE HEAT, TWO-STAGE COOL

FIGURE 18-21   COOL selected.

In Figure 18-23, electrical circuits are shown to both the heating and cooling sets of mercury bulbs. If the temperature at the thermostat is such that cooling is required, electrical circuits are completed to the Y_1 and/or Y_2 outputs, as necessary. On the other hand, if heating is required, electrical circuits will be completed to the W_1 and/or W_2 outputs, as necessary. Normal minimum separation between actuation of the heating and cooling mercury bulbs is about 4° F (2.2° C). This separation can be increased by the thermostat setting.

SIMPLIFIED SYSTEM SCHEMATIC: TWO-STAGE COOL, TWO-STAGE STRIP HEAT

The schematic of the control system of the two-stage cool, two-stage heat strip is shown in Figure 18-24. The operation of the system is controlled by low-voltage relays and contactors. The high-voltage heat strips, compressor motors, and inside fan motor are controlled only by their associated contactor or relay contacts.

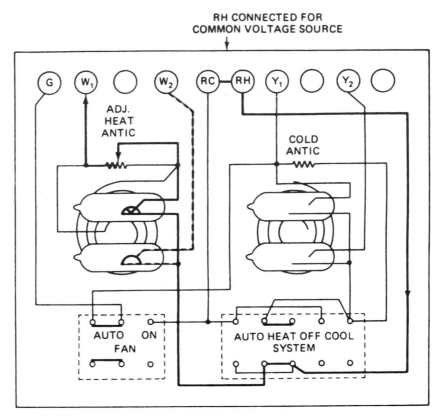

THERMOSTAT—TWO-STAGE HEAT, TWO-STAGE COOL

FIGURE 18-22 HEAT selected.

The heat strips have a thermal fuse link and high-temperature overload (cutout) connected in series with each strip. These are standard protective devices that are built into each heat strip assembly.

In the low-voltage circuits, the high-pressure and low-pressure control contacts are shown in series with compressor contactor coils only. Compressor 2 has an additional series control, a set of contacts in a device known as a time-delay relay. The circuit of the time-delay heater is from the Y_2 output of the thermostat. The opposite end of the time-delay heater is connected to the 24-volt return line. When voltage appears at Y_2 and after a predetermined time, the time-delay heater causes its contacts to close. A circuit to compressor contactor 2 is then complete through the high- and low-pressure controls.

The purpose of the time-delay relay is to ensure that both compressors do not start at the same time. In certain systems, the high starting current of two compressors starting simultaneously could overload the electrical power source.

The heat relay coil is connected in parallel with the coil of heat contactor 1. This relay is also used to provide a circuit to the inside blower motor contactor coil when heat is required. There is a circuit from the W_1 output of the

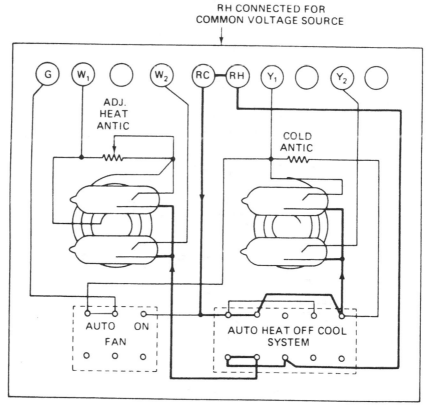

THERMOSTAT—TWO-STAGE HEAT, TWO-STAGE COOL

FIGURE 18-23   AUTO selected.

thermostat through the normally open contacts of the heat relay to the coil of the inside blower motor relay. The set of normally closed contacts of the heat relay is used to open the feedback circuit to the G terminal of the thermostat during heat operations.

WIRING DIAGRAM: TWO-STAGE HEAT, TWO-STAGE COOL

The wiring diagram in Figure 18-25 is of a two-stage heat, two-stage cool system. The wiring diagram differs from a schematic in that it shows the actual point of electrical connection of each wire. The wiring diagram shown in Figure 18-25 is the same circuit shown in Figure 18-24.

Cool Operation

Operation of the compressor motor and condenser fan motor is direct. When the associated contactor energizes, high-voltage power is fed to the compressor motor

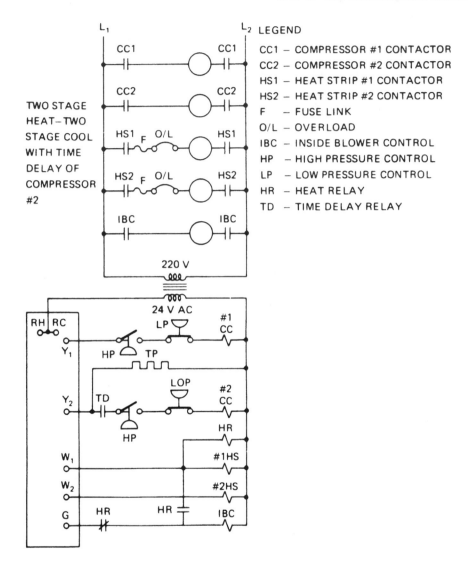

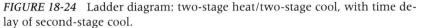

FIGURE 18-24 Ladder diagram: two-stage heat/two-stage cool, with time delay of second-stage cool.

and the condenser fan motor. The operation of the potential relay with associated start capacitor is as described in Unit 15.

The low-voltage control circuit for the condensing unit is shown in heavy lines in Figure 18-26. The 24 volts are available at the thermostat R terminal. They are also available at the Y_1 terminal when cooling is required. The circuit is completed from the Y_1 terminal through the pressure controls to the coil of contactor 1. The low-voltage return for the coil of contactor 1 is to the common of the 24-volt transformer. The wiring connection for the return wire is from contactor 1 via the contactor 2 coil common and then to the low-voltage return terminal strip.

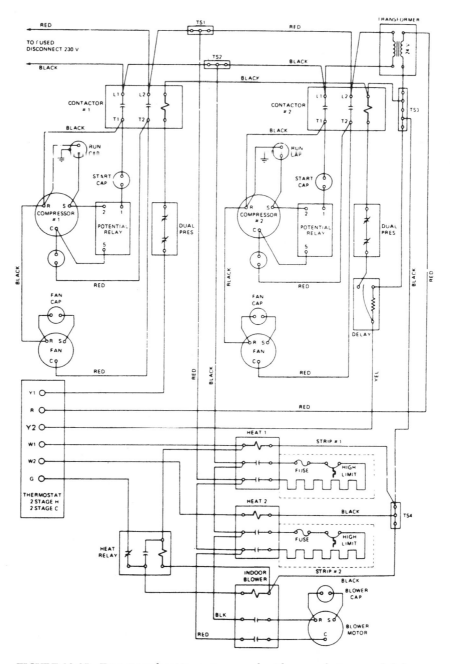

FIGURE 18-25 Two-stage heat/two-stage cool with second-stage cool delay.

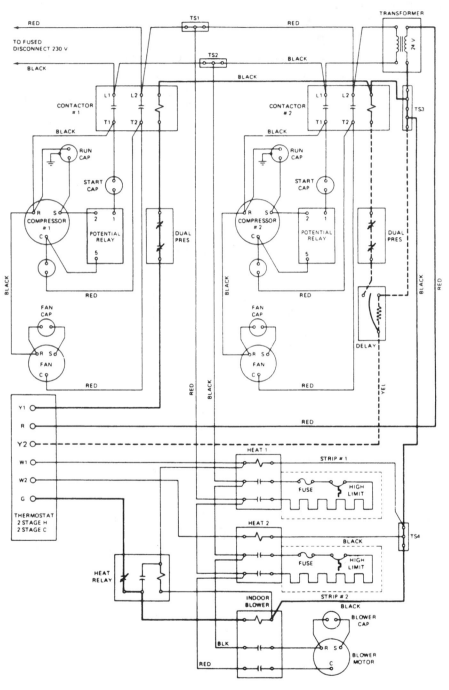

FIGURE 18-26   Cool operation

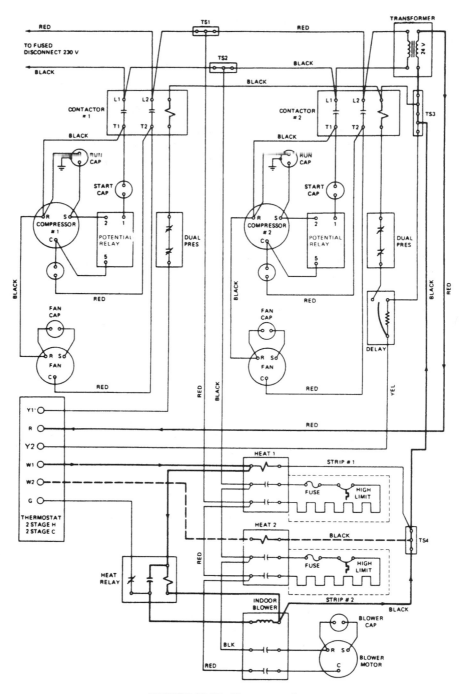

FIGURE 18-27 Heat operation

If two stages of cooling are required, then there must be 24 volts available at the Y_2 output of the thermostat. The heavy dashed line completes the circuit to the coil of contactor 2 through the time-delay relay and the dual-pressure control. The contacts of the time-delay relay close at a preset time period after power is available at Y_2. The return of the time-delay heater is to the return terminal strip.

The inside blower motor operates with control of the blower relay from the G output of the thermostat through the closed set of contacts of the heat relay to the coil of the blower contactor. The return side of the blower relay coil is connected to the 24-volt ac return.

Heat Operation

When heat is selected and required, 24 volts is available at the W_1 output of the thermostat. This voltage is fed to the coil of heat contactor 1 and to the coil of the heat relay, as shown in Figure 18-27.

The return circuit of heat contactor 1 coil and the heat relay coil are connected to the 24-volt return terminal strip. The strip is connected to the common of the 24-volt transformer.

The heat strip contactor and the heat relay energize. High-voltage power is fed through the heat strip contactor points to the strip. The heat strip is in the circuit, provided that the fuse link and overload are not open.

The 24-volt circuit is through the now closed heat relay contacts to the inside blower motor relay coil. The return side of this coil is connected back to the 24-volt common. The relay energizes, and high voltage is available to the blower motor.

The circuit to heat strip contactor 2 is shown in heavy dashed lines. Heat strip contactor 2 energizes whenever 24 volts is available at W_2 of the thermostat.

The normally closed set of contacts of the heat relay are now open, removing the feedback circuit to the G terminal of the thermostat. If this path is not opened and 24 volts made available at G of the thermostat, the compressor contactor would also become energized. This feedback circuit is shown in Figure 18-28.

The number of combinations of components used to provide air-conditioning circuits is as numerous as the number of air-conditioning manufacturers. It is important that technicians be familiar with the system used when working on an air-conditioning system. A wiring diagram is usually available; it is generally glued to the cover of the unit. It is sometimes practical to make a sketch or ladder diagram of the unit to facilitate troubleshooting.

DEHUMIDIFIERS

A system for controlled dehumidifying of air is rapidly becoming common in air-conditioning installations. Dehumidifying is a natural function of air conditioners in the cooling cycle, where water vapor is removed from the air as the air moves

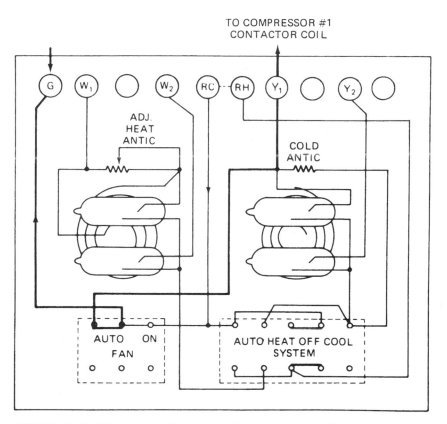

FIGURE 18-28   Thermostat–2-stage heat/2-heat cool. Feedback through thermostat if inside fan 24 V connection is not interrupted.

across the evaporator coil. Controlled dehumidifying is accomplished by reducing the speed of the evaporator blower under specific conditions during the cooling cycle. Controlled dehumidifying has become necessary since the advent of the new high-efficiency condensing units. The high-efficiency condensing units allow the evaporator to cool the air to the selected temperature so rapidly that little moisture is removed, and the humidity remains high. An early corrective procedure was to connect the evaporator blower for low-speed operation. This improved the dehumidifying capability of the system, but reduced its overall efficiency when high dehumidifying was not required.

The new controlled dehumidifying system uses a humidistat to sense the humidity level. During the cooling cycle, when the humidistat indicates humidity above the selected level, the evaporator blower will be operated at slow speed.

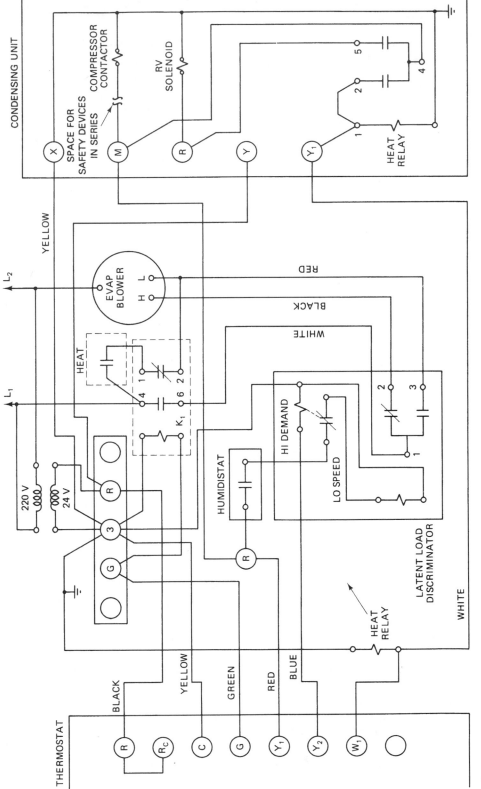

FIGURE 18-29 Humidity control.

When the humidistat senses humidity at or below the selected value, the evaporator will be switched back to high speed, providing maximum efficient cooling.

Circuit Components

Figure 18-29 is a simplified wiring diagram of an air-conditioning system including humidity control. Some control components have been left out of the drawing of the condenser unit. For example, the pressure controls and the anti-short-cycle controls are not shown. They complicate the drawing, but add nothing to the explanation of humidity control.

Locate the humidistat and latent load discriminator at the bottom center of the diagram. These components control the dehumidifying action of the evaporator during cooling.

Normal Cool and Heat Cycles

In the condensing unit, the RV solenoid is energized during the cooling cycle. Note the connection from M at the compressor contactor coil through the normally closed contacts, 4 and 5, of the heat relay to the RV solenoid input terminal R. When conditions call for cooling, there are 24 volts ac at the Y_1 terminal of the thermostat (cool selected and required). The compressor contactor and the RV solenoid will both energize.

When HEAT is selected, the heat relays will energize with power from thermostat W_1. Condenser heat relay contact 2 mates with 4, providing power to the compressor contactor input terminal M. The reverse cycle solenoid de-energizes as the heat relay energizes, and normally closed contacts 4 and 5 open. The system operates in reverse cycle, providing heat to the area.

Dehumidifying

Dehumidifying is accomplished by operating the evaporator blower at low speed during the cool cycle. This speed control comes about through the action of the humidistat and the latent load discriminator. The Y_1 output of the thermostat feeds 24 volts ac to the humidistat. If the humidity is above the selected value, the contacts of the humidistat will be closed. The 24-volt ac is fed through the humidistat and the normally closed contacts of the high-demand relay to the coil of the low-speed relay. The low-speed relay will energize, connecting line L1 to L, the low-speed terminal of the blower motor. Line L1 is connected through 4 and 6 contacts of K1, the blower relay. Relay K1 is energized from the G output of the thermostat. Under these conditions, the air-conditioning unit will operate in the cool cycle with the evaporator at low speed. Maximum moisture will be removed from the air.

If the demand for cooling is increased to the point where the thermostat provides for second-stage cooling, there will be 24 volts ac at Y_2. The high-demand relay in the latent load discriminator will energize. The normally closed contacts of the relay open, removing power from the coil of the low-speed relay. Contacts

1 and 2 of the low-speed relay mate, applying power to the high-speed input terminal of the blower motor. The evaporator blower would remain at high speed during periods of high demand for cooling. If cooling demand decreases, the 24-volt ac output at the thermostat, second-stage cooling, terminal Y_2, will be interrupted. The high-demand relay in the latent load discriminator de-energizes. Control of the speed of the evaporator blower motor returns to the humidistat.

SUMMARY

The review procedure for Unit 18 is to follow the circuits on each of the diagrams included. Be sure you understand how each component is energized.

PRACTICAL EXPERIENCE

Required equipment   A large heat pump with wiring diagrams.

Procedure
1. Locate each control and operating component of the system.
2. Compare the wiring diagram components to those located in the system.
3. Draw a ladder diagram of the system.
4. Which diagram is easier to follow when checking overall system operation?

Conclusion   It is important to have wiring and ladder diagrams of a system in order to check out system operation.

Refrigeration Circuits

OBJECTIVES

Upon completion and review of this unit, you will be familiar with refrigeration circuits, including the components that make up the circuits.

Many different circuits are used in the control of domestic refrigerators. In this unit, some of the most common circuits are covered. As with air-conditioning circuits, the best source of reference is the equipment manufacturers' service manuals and bulletins.

SIMPLE REFRIGERATOR

Figure 19-1 is a pictorial diagram of an electrical circuit for a basic refrigerator. The start relay is contained within the temperature-control (thermostat) unit. The same circuit is shown in schematic form in Figure 19-2. In the schematic, the start relay connections are more easily traced.

BASIC REFRIGERATOR/FREEZER COMBINATION

The circuit of a simple combination refrigerator/freezer is shown in Figure 19-3 in diagram form; operation of the defrost heater is not easily understood in the diagram. It is, however, more easily understood in the simplified schematic in Figure 19-4.

Whenever the temperature control cycles off, its contacts open, disconnecting the power circuit to the compressor. Current now flows through the defrost heater, compressor overload protector, start windings, and the start relay coil to the other side of the line. When the compressor is not running, the defrost heater is on, removing frost from the freezer coil section.

SIMPLE REFRIGERATOR

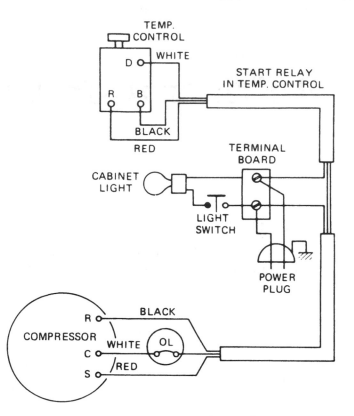

FIGURE 19-1   Wiring circuit, basic refrigerator.

MODERN, FROST-FREE REFRIGERATOR/FREEZER

As modern refrigerator-freezer technology has developed, requirements for frost and moisture control have increased. Combinations of heaters and timers are added to the basic circuits to provide the needed control.

The wiring layout (diagram) of a frost-free refrigerator is shown in Figure 19-5. The schematic of the same system is shown in Figure 19-6. Both the wiring layout and the schematic diagram are useful when servicing a unit. The wiring diagram provides wire color and point of connection information not included in the schematic diagram. The schematic provides a simplified relationship useful in troubleshooting. Together, the combination provides most of the required information for service work.

The operation of the unit is straightforward, with the refrigerator fan, door heater, mullion heater, defrost timer motor, and defrost heater connected in such a manner that power (115-volt ac) is applied whenever the plug is inserted in the wall socket. The refrigerator lamps light when the refrigerator door is open. The freezer lamp lights when the freezer door is open. In Figure 19-6 the associated

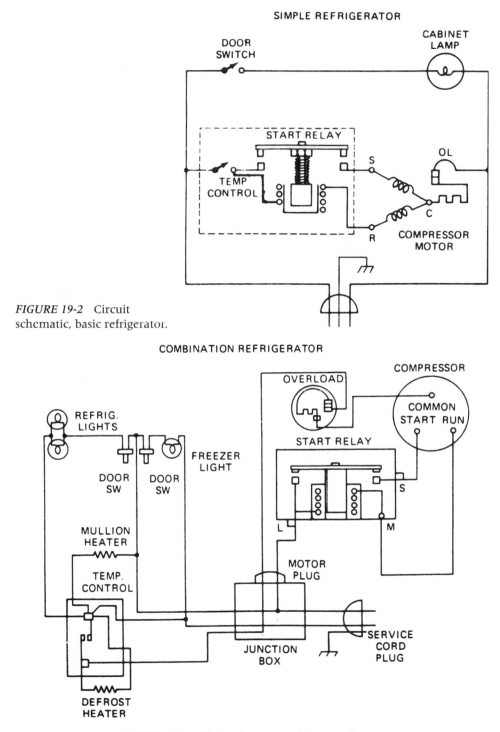

FIGURE 19-2 Circuit schematic, basic refrigerator.

FIGURE 19-3 Wiring diagram, refrigerator/freezer.

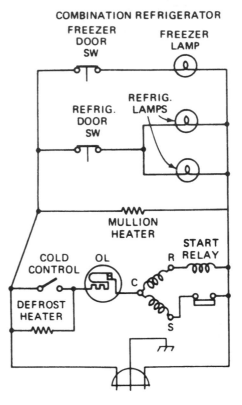

FIGURE 19-4 Simplified schematic, refrigerator/freezer.

door switches are shown in series with the lamps. The defrost timer is associated with the control of the freezer fan, compressor motor, drip (condensate) catcher, and defrost heaters.

The schedule of operation for this type of defrost timer varies with the manufacturer. One system design requires defrost heater operation for 17 minutes every 8 hours. The shaft of the timer motor is geared down to one revolution per 8 hours (3 rev/day). A cam on the shaft operates a switch that provides for the 17 minutes of defrost operation. The cam and switch arrangement is shown in Figure 19-7.

The normal circuit through the defrost timer contacts is from the input at terminal 3 through the switch to terminal 4 to the temperature control. The compressor motor and freezer fan motor operate as required when the thermostat control switch contacts are closed. The circuit through the defrost timer to the compressor is shown in Figure 19-8.

When the defrost timer motor cam actuates the switch, the compressor turns off even if the thermostat calls for cooling. The circuit to the defrost and drip-catcher heaters is completed for 17 minutes, as shown in Figure 19-9. The circuit through the defrost timer is from the input terminal 3 through the switch to the output terminal 2. The drip-catcher heater and defrost coil heater are connected from terminal 2 of the defrost timer to the other side of the line, as shown in Figure 19-6.

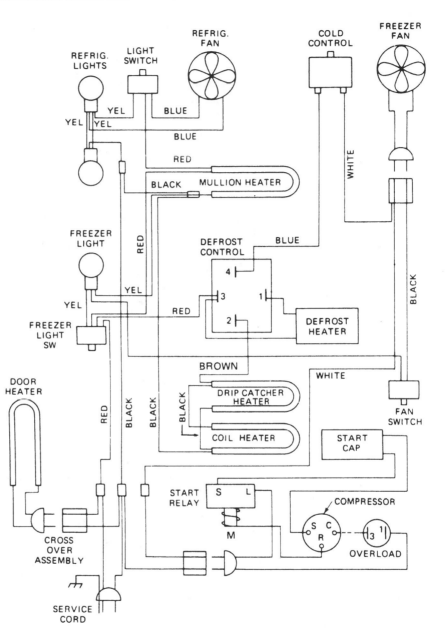

FIGURE 19-5   Wiring diagram of modern, frost-free refrigerator/freezer.

Accumulated Compressor Timer

The defrost timer in Figure 19-6 operates the defrost heater every 8 hours, whether the coil has frosted over or not. Actually, the amount of frost buildup depends on the total compressor operating time.

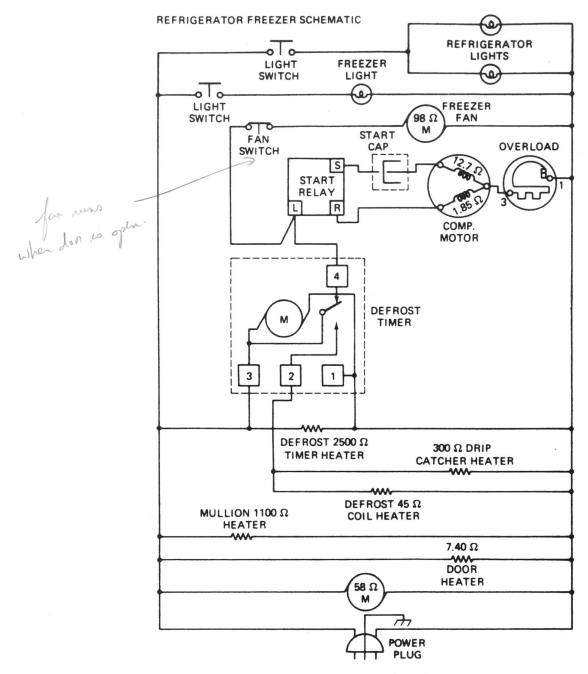

FIGURE 19-6 Refrigerator/freezer schematic.

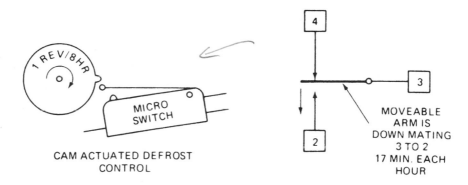

FIGURE 19-7   Cam and switch operation.

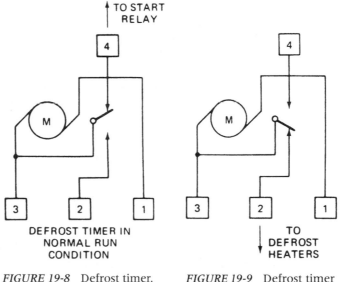

FIGURE 19-8   Defrost timer.   FIGURE 19-9   Defrost timer
                                    in defrost position.

The circuit in Figure 19-10 is that portion of a modern, frost-free refrigerator/ freezer associated with defrost. The defrost timer motor is controlled by the cold (thermostat) switch. The defrost timer motor rotates only when the cold control calls for compressor operation. After about 6 hours of accumulated compressor opera- tion, the defrost timer has rotated sufficiently to open the connection between ter- minals 3 and 4 of the timer. The compressor stops while the cold control continues to call for compressor operation. The circuit through the defrost timer is now ter- minals 3 to 2, and power is fed to the defrost heater. The switch remains in this po- sition for approximately 25 minutes. After this time period, the cam on the timer motor rotates past the timer switch, allowing power to be restored from terminal 3 to 4. The compressor operates and cools until the cold-control switch opens.

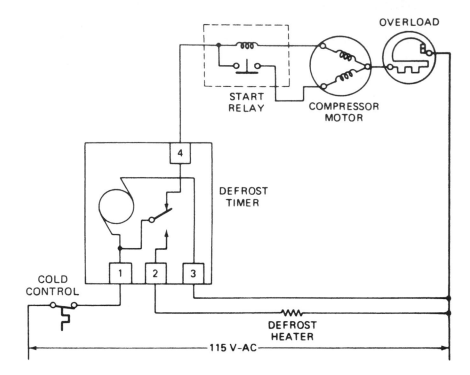

FIGURE 19-10   Accumulated compressor time.

Hot-Gas Defrost

The electrical diagram in Figure 19-11 is of a hot-gas defrost system with a defrost solenoid controlled by accumulated compressor time. The defrost timer motor is controlled by the cold-control thermostat. After sufficient compressor operating time, the cam on the defrost timer motor actuates the switch, causing power to be fed to the defrost solenoid. When the defrost solenoid is actuated, a metered amount of hot, compressed gas from the compressor is circulated through the evaporator unit, eliminating frost. Note that the compressor continues to operate during the defrost cycle.

As refrigerator/freezers developed, manufacturers added items, such as automatic ice makers; milk, juice, and soda dispensers; and cold-water faucets. The control circuits for these devices are as varied in design as there are manufacturers. Controls include heaters, switches, motors, and solenoids. It is necessary that the manufacturers' specifications and schematic drawings be referred to when these modern refrigerator/freezers are serviced.

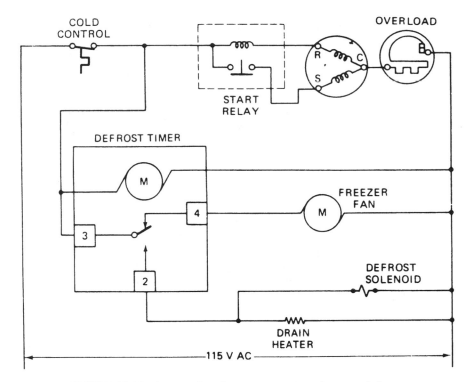

FIGURE 19-11   Accumulated compressor time, hot-gas defrost.

SUMMARY

As with the air conditioner, when learning system operation of the refrigerator, the wiring diagram and ladder diagram are important.

PRACTICAL EXPERIENCE

Required equipment   A refrigerator with automatic defrost, the wiring diagram of the system.

Procedure
1. Draw a ladder diagram of the system.
2. Locate the components of the refrigerator as shown on the diagram.

Unit 20

Troubleshooting

OBJECTIVE

Upon completion and review of this unit, you will be familiar with troubleshooting procedures.

Problems relating to an air-conditioning or refrigeration system malfunction can be divided into two categories: electrical and mechanical. It is estimated that about 80% of all service calls are because of electrical problems. Additionally, many mechanical problems become evident with the indication of an associated electrical problem.

Practical troubleshooting consists of relating the knowledge that the technician has to the problem to be solved. It is impractical to attempt to cover the infinite variety of possible troubles in a textbook. It is suggested that the technician concentrate on the basic theory of electricity, manufacturers' bulletins, controls, and system schematics of equipment.

Most air-conditioning and refrigeration systems consist of series circuits where individual controls are used to interrupt or complete electrical paths whenever a basic limitation is exceeded. The control operates to impose the limit on the circuit. It is the technician's responsibility to become familiar with the different control devices used in the related systems. Again, the manufacturers' bulletins, tech notes, and component information sheets provide the basic information on the operation of controls.

It is most important that the technician become familiar with electrical troubleshooting procedures. The following factors must be considered:

1. *Profit.* Actual time spent locating a defective part is just as important as (perhaps more important than) the time spent in making the repair or replacement. For example, if the problem is a defective fuse or circuit breaker, minimal time should be spent in determining the problem. The customer cannot afford—nor can the service company charge for—an excessive number of hours spent troubleshooting.

2. *Procedure.* It is important that a systematic procedure for troubleshooting be developed by the technician. By following a systematic procedure, the technician should be able to locate any problem in a minimum amount of time. For example, a defective major component (such as a compressor) should be detected almost as rapidly as a minor component (such as a fuse).

SYSTEMATIC TROUBLESHOOTING

Electrical troubleshooting of most air-conditioning systems may usually be divided into two subsystems: low voltage and high voltage. The low-voltage control circuit is usually 24 volts, supplied by a step-down transformer. The high-voltage power circuit is usually 220 volts, though some are 110, 115, 120, 208, 230, and even 440 volts. Steps involved are:

Localize When a problem exists in a system, the first procedure is to localize the problem. If the system contains sections, such as with a split air-conditioning system, the problem can usually be rapidly localized to a section, such as the evaporator or condenser section.

Isolate When troubleshooting, the technician should usually isolate the problem to a particular group or subsystem. Isolation of the problem usually leads to a more rapid discovery of the defective component.

Locate After the problem area is isolated to a specific section or system, the actual fault can be quickly located by following the suspect circuit to an open, short, or defective component. The final location of a single defect on an open or shorted component is not necessarily the end of the troubleshooting procedure. Often the defective part is merely an indication of a problem in another area. For example, a blown fuse or an open circuit breaker may indicate a shorted system component.

Conclude The conclusion of a problem is not accomplished until all defects relating to the problem are discovered and corrected. For example, an open high-pressure control indicates excessive high-side system pressure. If the cause of the excessive pressure is not corrected, such as a defective condenser fan motor, resetting the pressure control will be only a short-term cure of the complaint.

CUSTOMER COMPLAINTS

Often the information from the customer provides a lead to the trouble in a system. For example, when called upon to troubleshoot and repair a home air conditioner, 3-ton split system, the customer might comment: "No matter what I do with the thermostat, nothing happens. The air-conditioning unit doesn't make a sound." This is an indication that there could be a problem with either the high-voltage (220-volt) source or the low-voltage (24-volt) control. When the thermostat mode selections are varied, the sound of the high-voltage contactor energizing and de-energizing in the unit can usually be heard. If there is no high voltage, there may not be low voltage. Without low-voltage control, the contactor will not operate.

On another service call, the customer's comment may be: "When I adjust the thermostat, I can hear a click in the air conditioner, but the unit doesn't come on." This is an indication that 24-volt control power is available. It is possible that one side of the 220-volt source might be missing, provided that the step-down transformer is 110 volts to 24 volts. Of course, any number of other troubles are possible, but at least there is an indication that power is available.

Do not take the customer's indication of the trouble as being accurate. Good customer relations require that you listen to the comments. Then act according to established troubleshooting procedures.

LINE VOLTAGE

It is good practice to measure and record the line voltage present at the terminals of the power disconnect when on an air-conditioning service call. The line voltage greatly affects the operation of air-conditioning units. It should always be within the limits set by the manufacturer. The voltage should be recorded when on the service call for reference on a possible return call. Any change in normal voltage (higher or lower) is a possible trouble source. Also, be sure that the meter is accurate by checking it against a known value.

ELECTRICAL MEASUREMENTS IN TROUBLESHOOTING

For an air-conditioning system to function at full efficiency, the correct voltages must be present at the terminals of the motors used in the system.

In Figure 20-1, the schematic of a typical $1\frac{1}{2}$ ton, straight-cool window unit is shown. The problem given with the unit is: "When any COOL button is pressed, only the fan operates. There is no cool air output." These troubleshooting procedures should be followed:

1. Check the supply voltage at the wall plug to ensure that it is at the correct level as specified on the equipment nameplate. If the voltage is correct:

2. Connect the unit to the power source and press FAN. If the unit is multispeed, check each speed. If the fan operates properly:

3. Set the thermostat to maximum cool (HI-COOL), press one of the COOL buttons, and listen for compressor operation. Assuming that there is no action with the compressor, it will be necessary to remove the unit from its case to gain access to the electrical components. *Important:* When the unit is removed from the case, all power should be disconnected from the unit.

After the unit has been removed from the case, continue troubleshooting using a test lamp, voltmeter, or ohmmeter. A good method is to check the system with an ohmmeter, since line power is not needed for this instrument.

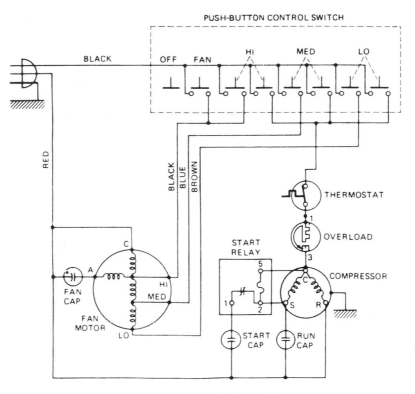

FIGURE 20-1 Window unit.

Since it has been determined that the compressor is not operating, the following fault possibilities should be considered:

1. No power to compressor motor (open circuit).

2. Defective component associated with compressor motor (capacitor or switch).

3. Defective compressor motor (shorted or open). The compressor motor may be mechanically seized. In this case, an open-short test would not indicate the problem.

Normally, if the compressor motor is shorted or open or if an associated component is defective, there are other indications, such as motor hum, when power is applied. Since no action was observed when the COOL button was depressed, the most logical procedure is to check for an open in the power circuit supplying the compressor motor.

In Figure 20-2, the schematic of the system is shown with an ohmmeter connected from the power input to the thermostat input, marked A. With the COOL button pressed, the ohmmeter should read 0 ohms. The ohmmeter scale should be $R \times 1$.

To check the thermostat, the ohmmeter wire at A should be moved to B. The ohmmeter reading should again be 0.

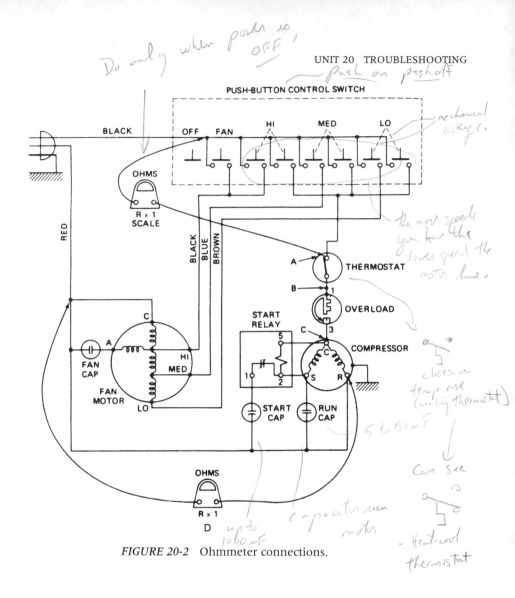

FIGURE 20-2 Ohmmeter connections.

To check the thermal overload, the meter lead at B should be moved to C. The ohmmeter reading should again be 0.

A defective component is indicated by a high resistance reading. If the ohmmeter reading is high when the lead is at A, the switch is defective. If the reading is 0 when at A but high when at B, the thermostat is defective. If the ohmmeter reading is zero at B, but high at C, the overload is defective. In this system, only the switch, thermostat, and overload are in series with the compressor motor. Connections and, especially, wire terminals of the push-on type must be checked. To check the compressor-motor return wire to the opposite power line, the ohmmeter may be connected as at D. Any resistance greater than 0 indicates a problem in this line.

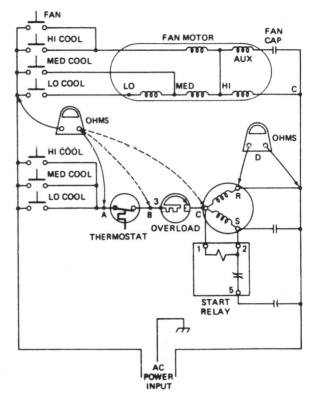

FIGURE 20-3 Ohmmeter connection, ladder diagram.

Figure 20-3 is a ladder diagram of the same system. Ohmmeter connections are shown with the same numbering system used in the schematic in Figure 20-2. Compare the schematic with the diagram. Comparing the two should provide for a better understanding of the procedures.

COMPRESSOR MOTOR PROBLEMS   *Troubleshooting*

Other kinds of problems in the compressor motor circuit provide different indications. If a motor winding is open or shorted or the other associated components are defective, there will be some sound from the compressor when the COOL button is first depressed. If something in the compressor motor circuit is defective and cannot start, there will be a heavy hum when power is applied. Power will then be interrupted by a blown fuse, opened circuit breaker, or the thermal overload. *or motor* If the thermal overload opens, it again closes (after a time period) and the heavy *may self* hum is again heard for a short time. *destruct.*

The following procedure may be used to check the associated components of a compressor motor.

Start Capacitor Test   A common trouble source with capacitor-start motors is the start capacitor. A simple test for an open or shorted capacitor may be made with an ohmmeter. The ohmmeter scale should be $R \times 100$ or $R \times 1000$. One side of the capacitor should have all wires removed from it.

The ohmmeter leads are connected across the capacitor terminals. A shorted capacitor is indicated by the meter pointer moving to 0 and remaining there. An open capacitor is indicated by no movement at all of the meter pointer. The proper indication for a sound capacitor on the meter is as follows: when the ohmmeter leads are connected to the capacitor, the pointer rapidly moves toward 0 and then slowly moves back toward maximum resistance. This happens because the capacitor charges to the meter's internal battery voltage.

Most start capacitor's have a 15,000-ohm, 2-watt resistor connected across the terminals. If the start capacitor does not have a resistor across its terminals, the capacitor must be discharged before attempting to check with an ohmmeter. Simply short the two capacitor terminals together, and then remove the short before making the resistance check with the ohmmeter. It is *not* necessary to remove this resistor before checking the capacitor. The ohmmeter pointer moves toward 0 and starts back up to the highest reading. The highest resistance that the meter will reach is 15,000 ohms, or the value of the resistor.

Compressor Test   Ohmmeter checkout of the compressor motor should be made with connecting leads removed from the motor terminals. With the ohmmeter on the lowest scale, $R \times 1$, the run winding measurement can be made with meter leads connected to terminals C and R. For a $1\frac{1}{2}$-ton compressor, the run winding resistance should be about 1.5 ohms. The start winding is checked with the meter leads connected to C and S terminals. The resistance of the start winding should be about 5 ohms. A check should also be made from the C terminal of the motor to the compressor case. This should provide an indication of an open or no circuit between the motor windings and the compressor case. The ohmmeter pointer should remain at the highest level when this measurement is made.

Start Relay Test (Potential Relay)   The start relay should be checked with all wires removed from terminals. With the ohmmeter in the $R \times 1$ range, connect the meter leads from terminals 1 and 2 of the relay. A completed circuit (0 ohms) should be indicated. The relay coil resistance measurement from terminals 2 and 5 should be above 1600 ohms in a 230-volt system.

Run Capacitor Test   The run capacitor is checked in the same manner as the start capacitor. The leads should be removed from one side of the capacitor. The terminals of the capacitor should be shorted together and then the short removed. Select a high-resistance range on the ohmmeter. When the meter leads are placed across the capacitor, the meter pointer should jump toward 0 and then rise toward the highest resistance reading. There is no bleed resistor on run capacitors.

FAN MOTOR PROBLEMS

— most common problem — fan gets dirty or blades get bent.

When the preliminary investigation of an air-conditioning system indicates that the fan motor is the trouble spot, troubleshooting should proceed in that portion of the system. With power disconnected, the fan motor shaft and fan blade should be rotated to ensure that an obstruction is not hindering rotation or that the motor bearings or bushings are not seized.

FAN MOTOR TESTING

The fan motor checkout procedure is similar to that used with the compressor motor. If the fan motor does not hum or give any indication of "power applied" when the compressor is operating, the problem is probably an open lead to the motor system. If the fan motor hums but does not rotate, the problem is probably due to an open or short in one of the motor windings. A defective fan capacitor may also cause the problem. If the fan motor runs properly at high speed but does not at medium speed, the problem is probably an open medium-speed winding or an open circuit to the medium-speed input terminal. If the fan motor runs properly at high and medium speeds, but not at low speed, the problem is probably in the low-speed winding or an open circuit to the low-speed input terminal.

Electrical Checkout Procedure

With the wall plug inserted in a checked 230-volt outlet, press the FAN button. If the fan operates properly, press the HI-COOL button and observe fan operation. If the fan rotates properly, press the MED-COOL button and observe fan operation. If the fan rotates properly, press the LO-COOL button.

If the fan does not operate properly in any or all speeds, the circuit should be tested using an ohmmeter. If the fan does not operate in any speed, disconnect the air-conditioner plug from the wall socket. Connect an ohmmeter (low-resistance scale) from the black lead input to the HI terminal of the fan motor (meter connection 1), as shown in Figure 20-4, and press the FAN button. The proper meter reading is 0 ohms. This checks the connection from the input power lead to the fan motor. If a complete circuit is indicated, connect the ohmmeter from the HI terminal of the fan motor to the C terminal of the fan motor (meter connection 2). This connection checks for an internal open in the motor. The HI, MED, and LO input leads should be disconnected during this ohmmeter check.

If the fan motor operates properly at HI speed but not at MED or LO speed, the motor lead should be checked out by pressing the appropriate fan speed button and connecting the ohmmeter leads from the input to the fan motor MED or LO speed terminal, as required.

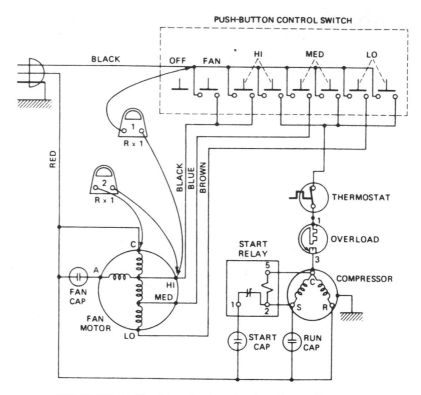

FIGURE 20-4   Checking the fan circuit with an ohmmeter.

To check for an open winding, the ohmmeter connection at the fan motor may be made as shown in Figure 20-5. The HI, MED, and LO motor leads should be disconnected when these ohmmeter checks are made. Connect the meter as to A to check the medium-speed winding and as at B to check the low-speed winding.

The auxiliary winding may be checked with the ohmmeter connected as shown at A in Figure 20-6. The isolation of motor windings from the motor frame

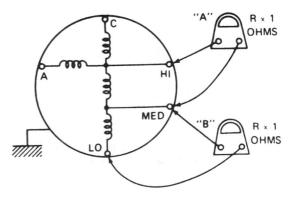

FIGURE 20-5   Checking the medium and low windings with an ohmmeter.

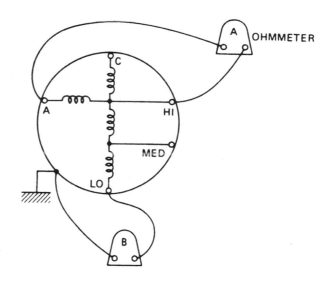

FIGURE 20-6 Checking the fan
with an ohmmeter.

may be checked by an ohmmeter connection from any input motor terminal to
the motor frame, as shown at B. The ohmmeter should be set on a high-resistance
scale for this procedure.

The fan capacitor may be checked out by the same procedure used with the
compressor motor capacitors. The leads are removed from one side of the capacitor. The ohmmeter is set on a scale of $R \times 1000$ or higher and is connected across
the capacitor. A good capacitor is indicated by the meter pointer jumping toward
0 and then rising toward the highest resistance reading. The connections are
shown in Figure 20-7.

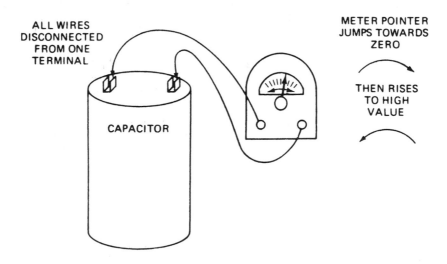

FIGURE 20-7 Fan capacitor check using ohmmeter.

SPLIT SYSTEM: SINGLE-STAGE, REVERSE CYCLE WITH HEAT STRIP

The schematic of a split air-conditioning system is shown in Figure 20-8. The troubleshooting procedures used in this unit are somewhat different from those with window or through-the-wall units.

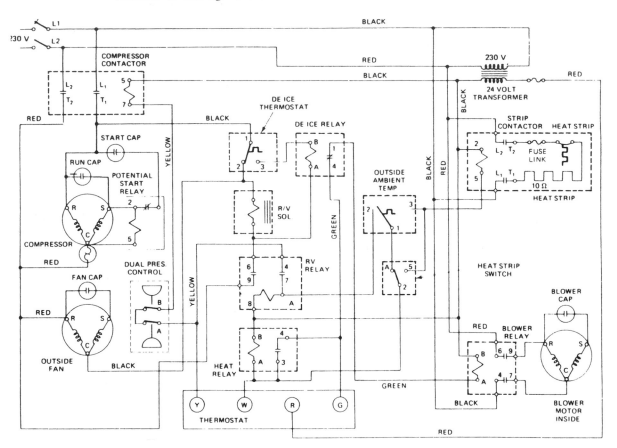

FIGURE 20-8   Split system: single-stage reverse cycle with heat strip.

In this split system, there are two electrical systems: 230-volt (power) high voltage and the 24-volt (control) low voltage. Troubleshooting electrical measurements is usually made in sequence, in one system at a time until an indication of the problem is found.

Localize the Problem

When troubleshooting the larger systems, it is important to follow the systematic approach: localize, isolate, and locate. It is *never* a good procedure to make haphazard measurements in a system, hoping that luck will lead to the trouble spot.

As with smaller units, the customer's complaint is often sufficient to localize the trouble. In some instances, however, preliminary procedures must be followed to determine which section of the unit is malfunctioning. For example, the customer's complaint might be useful in localizing a problem to the inside blower motor when the complaint is given that no air comes out of the vent when COOL is selected and room temperature is above the selected temperature. Proper procedure calls for checking the thermostat settings. The fan AUTO-ON switch should be moved back and forth to determine whether the inside fan operates in either position. The operation of the outside condenser system should also be checked. If the inside blower motor does not operate when the rest of the system does, the problem has been *localized*.

Isolate the Problem

After the problem has been localized to a section of the air-conditioning unit, it should be isolated to a particular subsystem. To isolate the problem, the technician might make a few voltage checks to determine the presence of high voltage between terminals 6 and 4 of the blower relay. Voltmeter connections are shown in Figure 20-9. If the proper 230 volts is present between terminals 6 and 4 of the blower relay, the meter leads should be moved to terminals 9 and 7 of the relay. If there is no voltage between terminals 9 and 7, a further check should be made to *isolate* the trouble. The relay coil voltage should be measured, with voltmeter connections as shown (meter 2). If 24 volts is present across the relay coil, the problem has been *isolated*.

Locate the Problem

With this particular problem, isolation and location are essentially the same thing. Normal procedure calls for replacement of the relay.

To replace the relay, *turn off the power with the main power switch*. Then give the relay a good visual inspection before removal, to ensure that there are no physical obstructions to the contact movement. The relay coil should be checked for an open with an ohmmeter; if the coil is found open, the problem has been located, and replacement can be accomplished with assurance that the trouble has been eliminated. To feel secure that a problem has been eliminated, it is necessary to *locate* the trouble.

In Figure 20-10, a portion of the air-conditioning system is shown in ladder form. In the previous paragraphs, measurements indicated that 230 volts was present between terminals 4 and 6 of the relay. If the voltmeter had not indicated 24 volts ac at position 3, a different problem would exist, and further troubleshooting would be required to isolate and locate trouble.

Procedure

If the voltage indication of voltmeter position 3 is 0 volts, an open must exist in the 24-volt circuit. A voltmeter check of the low-voltage transformer must be

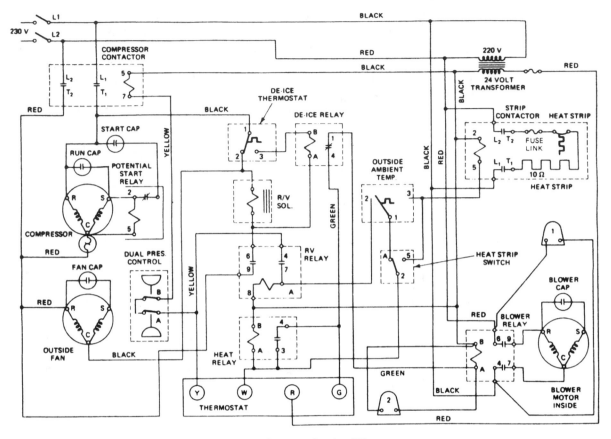

FIGURE 20-9   Voltmeter check of blower motor.

made. This measurement is taken at the transformer secondary, as shown by meter position 4. The low-voltage fuse (if equipped) is checked by moving one of the meter leads, shown as meter position 5.

De-ice relay contacts may be checked with the voltmeter connected as shown by voltmeter position 6. The meter should indicate 0 volts if the de-ice relay contacts are closed. A reading of 24 volts ac indicates an open circuit in the de-ice thermostat.

The thermostat may be checked with the voltmeter connected as shown by position 7 from R to G. Again, the voltmeter should indicate 0 volts since the circuit should be complete through the thermostat. A voltmeter reading of 24 volts ac from R to G indicates an open within the thermostat. If the problem is isolated to the thermostat, the main power switch to the system should be opened. Further checks may be made with an ohmmeter to *locate* the trouble.

Remember to lock the switch in the open position.

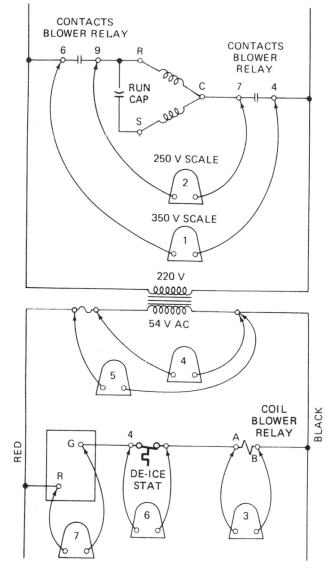

FIGURE 20-10 Voltage check-points, ladder diagram, dual-voltage system.

When attempting to *locate* a problem in a thermostat, visual inspection is of great importance. The problem of loose wires to connections must be considered. Check and tighten terminal connections, and then recheck for proper operation. The circuit through the thermostat from the R input to the G output may be followed using a voltmeter with power on or an ohmmeter with power off. A diagram of a thermostat for the split system is shown in Figure 20-11.

When a problem is *isolated* to a thermostat, the thermostat may be replaced and the air-conditioning system returned to operation.

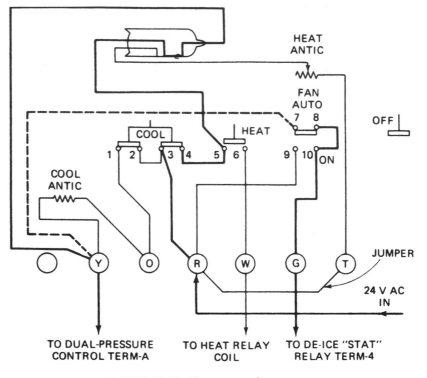

FIGURE 20-11 Thermostat diagram.

When possible, the isolated trouble should be located within the device. Sometimes the part that fails is a symptom of a more important problem area. When a defective part is located, it is usually possible to determine what caused the trouble.

It is always a good practice to eliminate the cause, rather than the results, of a trouble. However, the service technician must do repair work in the shortest time possible. If a repeat call is made on a unit and the part that was originally replaced is again malfunctioning, it is reasonable to suspect that the malfunctioning part is a symptom rather than the total problem. In almost every repair situation, a judgment has to be made whether to investigate further in a problem area or only to replace the defective part.

Power-On Check In this example, the problem has been isolated to the thermostat. With proper control settings, COOL selected and fan in AUTO, and room temperature above selected temperature, the inside blower motor does not operate. A voltage check of the thermostat indicated 24 volts ac between R and G of the thermostat. Voltmeter connections are shown in Figure 20-12.

Since the compressor system is operating, the circuit through the thermostat R to Y is proper. The circuit from Y to G, shown in a heavy dashed line in Figure

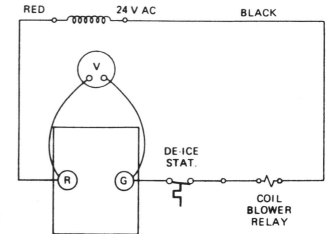

FIGURE 20-12 Thermostat check with a voltmeter.

20-11, is not complete. The fan switch should be placed in the ON position to isolate the problem further. If the inside blower now operates, the problem has been isolated to the auto section.

If the voltmeter indicates 24 volts between R and G of the thermostat when the fan switch is in either AUTO or ON, the problem has been isolated to the fan switch. If the terminals of the fan switch are accessible, the voltmeter leads might be placed directly across switch terminals 7 and 8 with the fan in AUTO. A reading of 24 volts indicates an open circuit across a set of switch contacts that is supposed to be closed. The switch is defective.

Power-Off Check After the problem has been isolated to the thermostat, the device may be checked out with power turned off. The main power disconnect switch should be *locked* in the off position.

With the conditions previously given, there should be a complete circuit between terminals R and G of the thermostat. An ohmmeter ($R \times 1$ scale) connected to R and G should, under normal conditions, indicate 0 ohms. If the fan switch is open as was previously indicated, the internal resistance of the thermostat between R and G will be extremely high. The ohmmeter reading will approach infinity.

If, in this test, the external connections are not removed from terminals R and G of the thermostat, the ohmmeter will indicate the resistance of the external circuit. Total resistance is that of the de-ice thermostat (very low), the blower relay coil, and the 240/24-volt transformer secondary, as shown in Figure 20-13.

To check the thermostat without external resistance loads, either the input lead at terminal R or the output lead at terminal G must be disconnected. The ohmmeter then indicates the resistance through the thermostat circuit only, as shown in Figure 20-14.

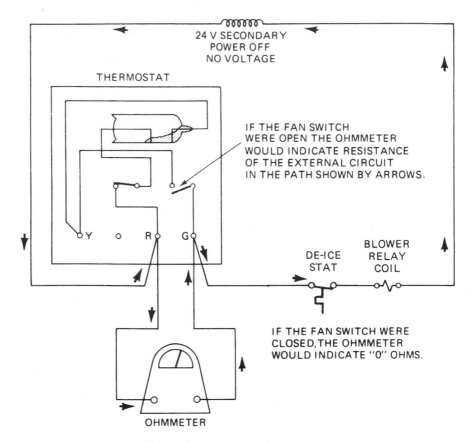

FIGURE 20-13   Thermostat check with an ohmmeter.

REFRIGERATION CIRCUITS

The schematics and wiring diagrams are very important to the service technician when troubleshooting a refrigeration system. Because there are many variations of wiring systems, troubleshooting is always easier when tracing a circuit in a schematic. As with air-conditioning systems, the refrigerator electrical system consists mainly of a few operating parts in parallel with control devices in series.

A wiring diagram of a typical refrigerator is shown in Figure 20-15. A schematic of the same unit is shown in Figure 20-16. A service call complaint may be that the lights and fan work, but the freezer and refrigerator are warm. Troubleshooting procedures for this complaint are as follows.

Localize

Since the lights and fan work, it is reasonable to suspect that the problem lies with the compressor circuit. However, to ensure that line voltage is adequate, a voltage check may be made at the power source, the wall outlet.

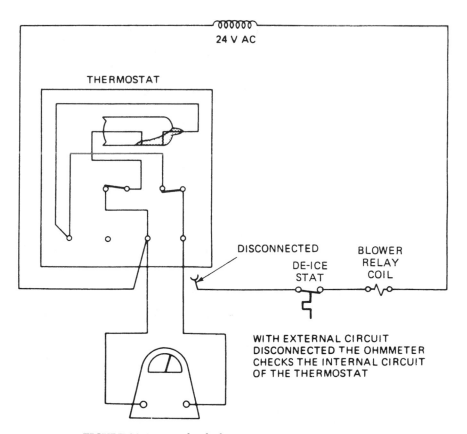

24 V AC

THERMOSTAT

DISCONNECTED

DE-ICE
STAT

BLOWER
RELAY
COIL

WITH EXTERNAL CIRCUIT
DISCONNECTED THE OHMMETER
CHECKS THE INTERNAL CIRCUIT
OF THE THERMOSTAT

FIGURE 20-14   To check thermostat, remove external load.

Isolate

Be sure that the thermostat (cold control) is adjusted to its coldest position. If the compressor motor still does not operate, disconnect the wall plug and then the compressor assembly disconnect plug.

An ohmmeter check of the compressor assembly can be made by connecting an ohmmeter across the terminals of the disconnected (compressor assembly) plug. The resistance reading should be about 2 ohms, the resistance of the run winding. As can be noted in Figure 20-16, the circuit to the start winding is interrupted by the start relay. The overload should have minimum (0) resistance.

If the compressor circuit checks out properly, the remainder of the circuit may be checked from the same point. The compressor assembly disconnect plug should have a complete circuit back to the input service plug. Ohmmeter checks may be made to determine if complete circuits exist.

If a complete circuit is not shown in the first check from the service cord to the compressor assembly disconnect, a connection should be made to the other side of the service cord plug. Ohmmeter connections are shown as 1 and 2 in Figure 20-17. The paths to complete the circuit (for 0 resistance) are also shown.

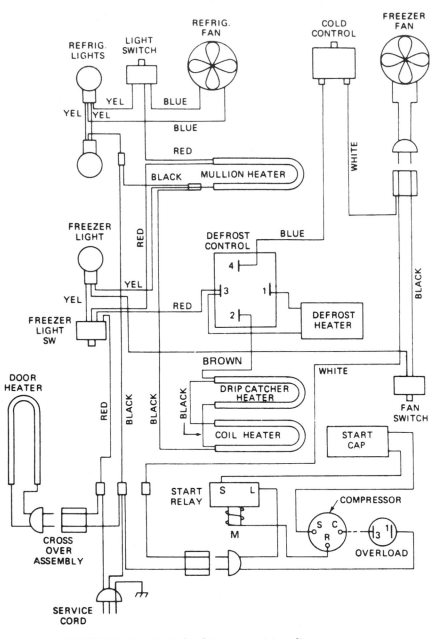

FIGURE 20-15 Typical refrigerator wiring diagram.

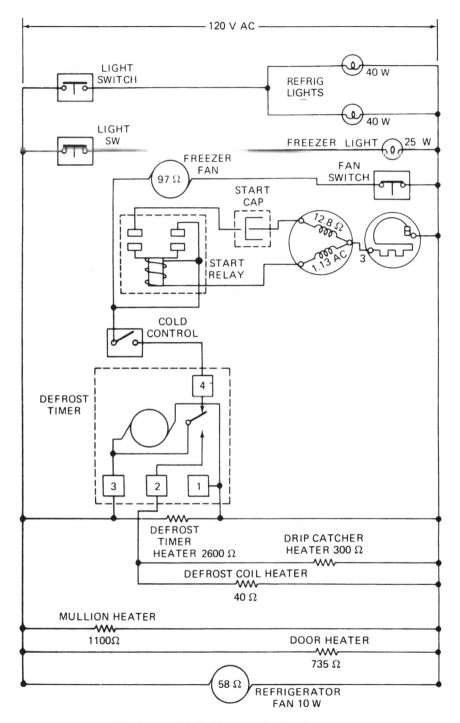

FIGURE 20-16 Refrigerator ladder diagram.

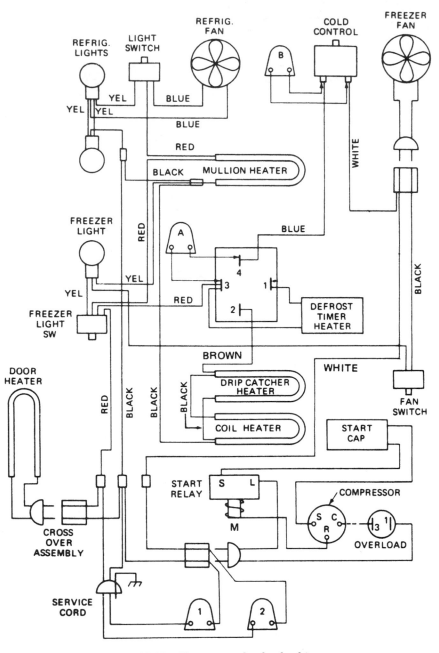

FIGURE 20-17   Ohmmeter check of refrigerator.

   If the 1 connection indicates 0 ohms but the 2 connection indicates an open (∞ ohms), the problem has been isolated to the control circuit to the compressor. The control circuit consists of two switches, the defrost timer and the cold control, in series with the interconnecting wires.

Locate

   An ohmmeter check may be made between terminals 3 and 4 of the defrost timer. The resistance should be 0 ohms. An ohmmeter check may also be made of the cold control. The resistance should be 0 ohms. Connections are shown in Figure 20-17. If both of these indications are proper, an open in the wiring is suspected. Remember that the problem has been isolated to the circuit between the service-cord plug connection (on the left) and the compressor-assembly disconnect jack (on the top). Checking the wiring requires following the circuit.

   An ohmmeter connection should be made to the service cord pin connection on the left and then down through connections 1 through 4, until the open circuit or component is located (Figure 20-18). The ohmmeter indication should be 0 ohms with each connection until the open is reached. When the open is discovered, a visual inspection will usually locate the problem. In this case, there is probably a broken wire or loose connection at a terminal.

POWER-ON CHECKS

   The same checkout sequence may be made using a voltmeter or test lamp. A low-wattage test lamp is preferred, usually 60 watts or less. The sequence is shown in Figure 20-19. The voltmeter indicates 0 volts across a closed circuit and 120 volts across an open circuit. The test lamp does not light across a closed circuit, but it does light across an open circuit.

   Remember, troubleshooting in air-conditioning and refrigeration circuits is best done in a standard sequence.

1. Listen carefully to the customer's complaint for clues to the trouble area.
2. Make a good visual inspection of the equipment for obvious troubles.
3. Review the schematics and wiring diagrams for clues to probable trouble spots.
4. Localize, isolate, and locate the trouble.
5. Return the device to proper working order in as short a period of time as possible. Do not, however, sacrifice time for accuracy and completeness.
6. Practice safe habits. Remember that getting too friendly with electricity can be a *shocking* experience.

SUMMARY

In troubleshooting, the proper procedure is: localize, isolate, and locate.

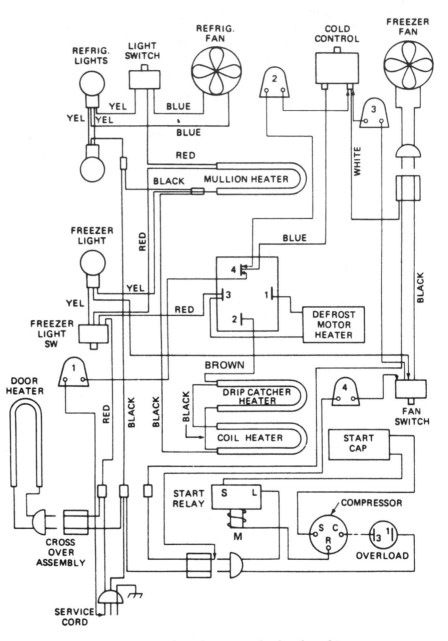

FIGURE 20-18   Further ohmmeter checks of a refrigerator.

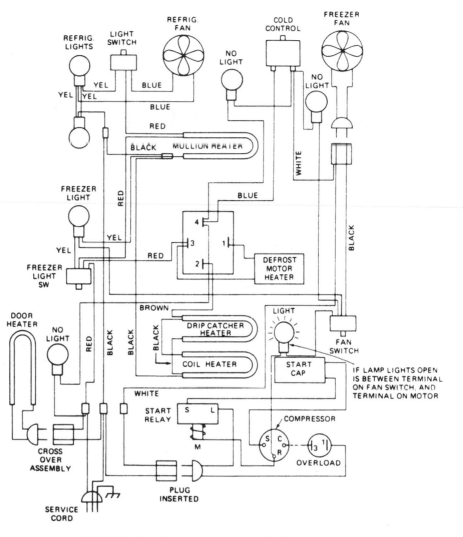

FIGURE 20-19   Using a test lamp to find an open circuit.

PRACTICAL EXPERIENCE

Required equipment   A malfunctioning air-conditioning system.

Procedure

1. Using the standard troubleshooting procedure:
   a. Localize the problem.
   b. Isolate the problem.
   c. Locate the problem.
2. Replace the malfunctioning devices, returning the unit to normal operation.

Unit **21**

Gas-Furnace Controls

OBJECTIVE

Upon completion and review of this unit, you will understand gas-furnace controls.

Gas-furnace control has been developed to a level of safety that both the furnace manufacturing and service industries may be proud of. The standard controls associated with the flow of gas have contributed to establishment of an excellent record of service. These controls include

1. Pressure regulators.
2. Gas valves (electric).
3. Thermocouples.
4. Thermopiles (power thermocouples).
5. Thermopilot relays.
6. Limit controls.
7. Thermostats.
8. High-voltage ignitor transformers.

STANDARD SYSTEM

A common system includes a thermopilot valve, main solenoid gas valve, thermocouple, thermostat, and limit control. A sample system of gas furnace control is shown in Figure 21-1.

The thermopilot valve shown in Figure 21-2 controls the flow of gas to the pilot and to the main burner solenoid gas valve. The solenoid gas valve used in this system for main burner control is shown in Figure 21-3. Spring action keeps this valve closed when power is not supplied to the solenoid coil. When power is supplied to the solenoid coil, the valve opens, allowing gas to flow as shown in Figure 21-3b.

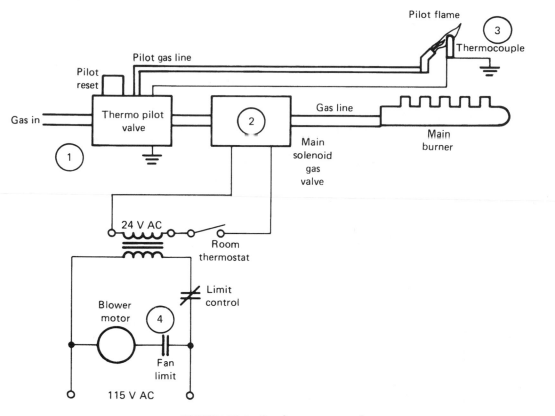

FIGURE 21-1   Gas furnace control system.

The thermopilot valve, numbered 1 in Figure 21-1, controls gas flow to the pilot burner and to the main burner's solenoid gas valve. The pilot may be lit when the reset button is pressed. Pressing the reset button allows gas to the pilot burner only.

The pilot burner, when lit, heats the thermocouple, numbered 3 in Figure 21-1. The heated thermocouple supplies thermoelectric current to the solenoid coil in the thermopilot valve. When the thermocouple is heated to a sufficiently high temperature by the pilot light, it will provide enough current to keep the solenoid coil in the thermopilot valve energized.

When this solenoid coil is energized, the release of the reset button allows gas to continue to flow to the pilot burner. Gas is also allowed to flow to the main burner's solenoid gas valve, numbered 2 in Figure 21-1. The main burner's solenoid gas valve is controlled by the thermostat. When the thermostat calls for heat, 24 volts ac is fed to the solenoid gas valve. The solenoid coil is activated, opening the valve and allowing the flow of gas to the main burner. Since the pilot is already lit, the main burner is ignited by the pilot.

The indoor blower operates after the plenum thermostat closes, shown as 4 in Figure 21-1. This thermostat, often called a fan limit, closes when the air in the plenum is heated, thus keeping the blower from moving cold air into the space to

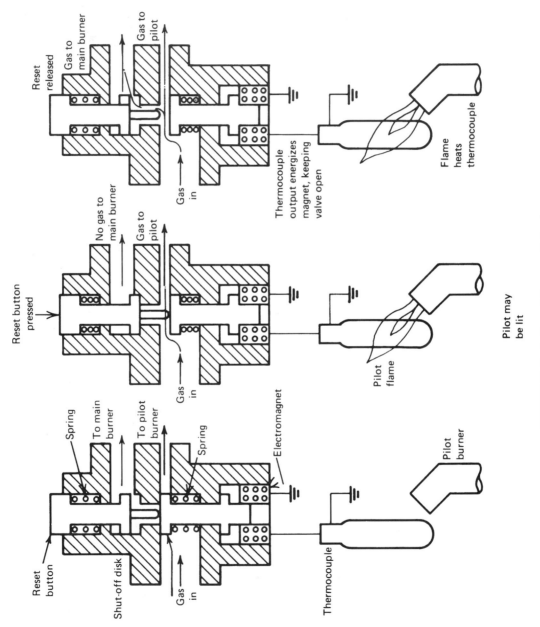

FIGURE 21-2   Action of thermopilot valve.

310

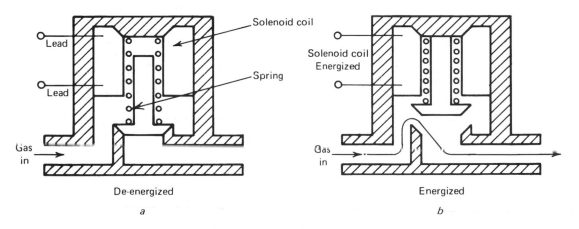

FIGURE 21-3 Solenoid gas valve.

be heated. A limit thermostat also senses plenum heat. If the temperature in the plenum becomes too high, the limit thermostat will open, removing primary voltage from the control transformer primary. Without primary voltage, there is no secondary control voltage. The main burner solenoid gas valve de-energizes and closes, cutting off the flow of gas to the main burner. The system is automatically reset when the plenum temperature is reduced to a level below the limit thermostat setting. Normally, the limit thermostat would operate only if the blower system should fail.

100% SHUT-OFF SYSTEM

A further safety feature may be included in gas heating systems by providing 100% gas shutoff if a limit is exceeded. This system is shown in Figure 21-4. The limit control is connected in series with the thermostat solenoid valve coil. The limit control, when open, will interrupt the circuit from the thermocouple. The thermopilot solenoid gas valve will de-energize and close. This shuts off gas to both the pilot and the main burner.

100% SHUT-OFF MILLIVOLT SYSTEM

Another safety feature is the two-valve millivolt system. In this system, both the thermopilot solenoid gas valve and main gas valve operate from the thermocouple current. The millivolt circuit is shown in Figure 21-5. If for any reason a limit is reached, the circuit to the thermopilot solenoid gas valve will be interrupted. When the thermopilot solenoid coil de-energizes, the valve will close, shutting off gas flow to both the pilot and the main burner gas lines. When the pilot goes off, the thermocouple will cool down. The decreased electric output from the thermocouple will cause the main burner solenoid coil to drop out.

The thermocouple used in this system may be a "power pile," which is nothing more than a series of thermocouple junctions connected in a series circuit. A

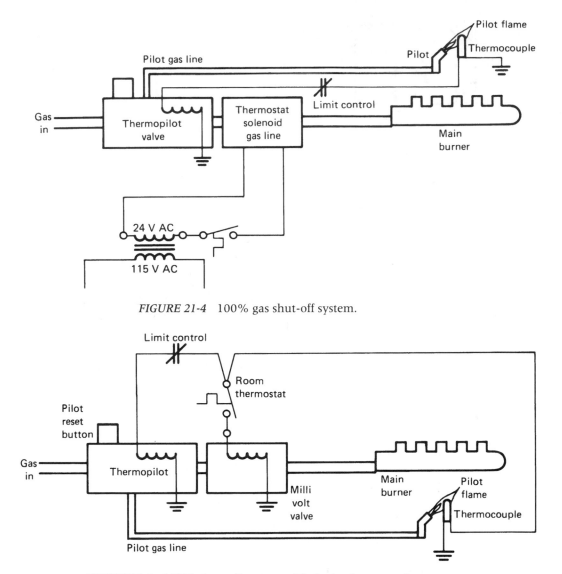

FIGURE 21-4   100% gas shut-off system.

FIGURE 21-5   100% shut-off system with dual valve control.

single thermocouple will produce 30 millivolts across its output. Power piles may produce from 250 millivolts to 750 millivolts.

ENERGY-SAVING INTERMITTENT-PILOT IGNITION SYSTEM

The new intermittent pilot gas systems are coming into general use as a means of conserving energy. The intermittent-pilot gas system uses gas only when the system calls for heat. This system eliminates the continuously burning gas pilot light. The components that make up the system are

1. Heat thermostat.
2. Time delay.
3. Buzzer.
4. Plenum temperature limit.
5. Capacitive discharge ignition.
6. Ignition electrode.
7. Thermocouple.
8. Millivolt relay.
9. Pilot gas valve with gas pressure switch.
10. Main gas valve.

> NOTE: *Figure 21-6 shows the general arrangement of components making up an energy-saving system. The colors given for color coded wires are used as a reference in this drawing only. They are not to be considered standard for energy-saving systems. The drawing is not meant to be a copy of any manufacturer's system. The individual manufacturer's service bulletin should be referred to when servicing equipment.*

The control transformer (1) in Figure 21-6 provides 24 volts ac for operation of the energy-saving system. The red wire output from the transformer feeds into the room thermostat (2). When the thermostat calls for heat, the 24 volts is fed out of the thermostat on the white wire to the time delay (3). The circuit continues through the normally closed contacts of the time delay to the capacitive discharge ignition unit (4). The return side (black wire) of the capacitive discharge ignition unit is connected through the normally closed contacts 2 and 1 of the millivolt relay (5) to the control transformer. When power is fed to the capacitive discharge ignition unit, it produces a high-voltage output. The high voltage is fed to the electrodes (11), which emit sparks to ground.

The 24 volts is at the same time fed from the input of the capacitive discharge ignition unit on the white/red wire to the main gas solenoid valve (9) and the pilot gas solenoid valve (8). The return side of the pilot gas solenoid valve is permanently connected to the return side of the 24-volt control transformer with a black wire. The pilot gas solenoid valve will energize, providing gas to the pilot and to the input of the main gas solenoid valve. A pressure switch (12) in the pilot gas valve will close if the gas pressure is high enough. The pilot will light with gas from the pilot gas solenoid valve and spark from the capacitive discharge ignition system. The flame from the pilot light will heat the thermocouple (7). The low-voltage output of the thermocouple is fed through the gas pressure switch (12) to the millivolt relay. When the thermocouple is sufficiently hot, the millivolt relay will energize, breaking the 1 and 2 contacts and making the 1 and 3 contacts. The capacitive discharge ignition will stop producing high voltage and the sparking at the electrodes will stop. With the 1 and 3 terminals of the millivolt relay making contact, the main solenoid gas valve has a return connection to the 24-volt control transformer. The main gas solenoid valve opens, and the main burner lights because the pilot flame is producing the required heat. When the building is heated, the room thermostat contacts will open. This removes the 24

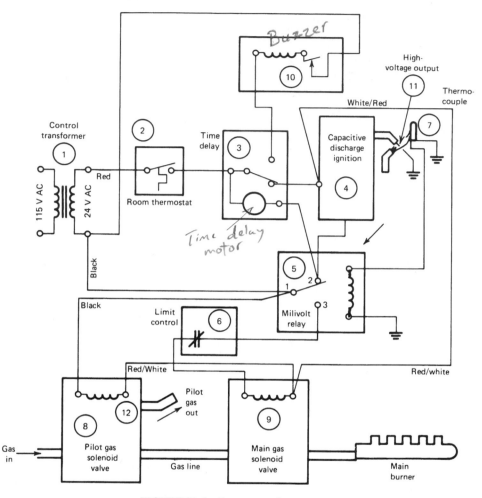

FIGURE 21-6 Energy-saving system.

volts from the system. Both the pilot gas valve and the main gas valve will close, shutting off all gas flow.

SAFETY FEATURES

The plenum temperature limit (6) provides the same protection as it does when used with the standard gas heat systems. If, for any reason, the plenum temperature should rise above the preset limit, the contacts of the limit control would open the system. The main gas valves would close, shutting off the flow of gas to the main burner.

The gas pressure switch (12) is located in the housing of the pilot gas valve. If gas pressure should drop, the gas pressure switch would open. The millivolt relay would de-energize, removing the connection from 1 to 3 of the relay. This would interrupt the return line connection from the main gas valve and the valve would close, shutting off the flow of gas to the main burner.

The electrodes of the high-voltage ignition system would immediately start to spark at the ignition electrode in order to restart or ensure continuous flow at the pilot light. When the gas pressure returned to normal, the pilot would light, heating the thermocouple, or if the pilot had remained lit, the millivolt relay would energize when the gas pressure switch reclosed. The high-voltage ignition would again cut off, and the main gas valve would again open.

If, for any reason, the millivolt relay does not energize after the room thermostat closes, the time delay will open the electric circuit to the system after 90 seconds. If the time-delay motor (3) is allowed to run for 90 seconds, a buzzer (1) will sound, indicating a problem. Under normal conditions, the time-delay motor only runs for a few seconds as the thermocouple heats, that is, until the thermocouple provides sufficient voltage to energize the millivolt relay. The return line connection for the time-delay motor is through the 2 and 1 contacts of the millivolt relay. This circuit is opened shortly after the pilot is lit. The time-delay motor is spring-loaded to return to zero.

MODIFIED SYSTEMS

Some of the latest energy-saving systems use electronic time-delay relays. The device is not repairable and must be replaced if it malfunctions. The connections to the electronic time delay are shown in Figure 21-7. The unit can be used as a direct replacement for the mechanical time delay shown in Figure 21-6.

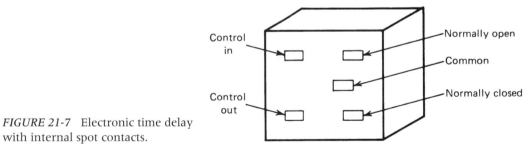

FIGURE 21-7 Electronic time delay with internal spot contacts.

An electronic time delay with a single set of normally open contacts could also be used, along with an external relay in the circuit. The electronic time delay would only control the external relay, which would perform the transfer function in the circuit. An example is shown in Figure 21-8.

Two recent developments are the hydraulic flame sensor and the electronic flame sensor.
⤷ book published in 1997.
       last

Hydraulic Flame Sensor

The hydraulic flame sensor uses a bulb and capillary assembly to control an electric switch. When the pilot flame heats the bulb, the fluid in the bulb expands,

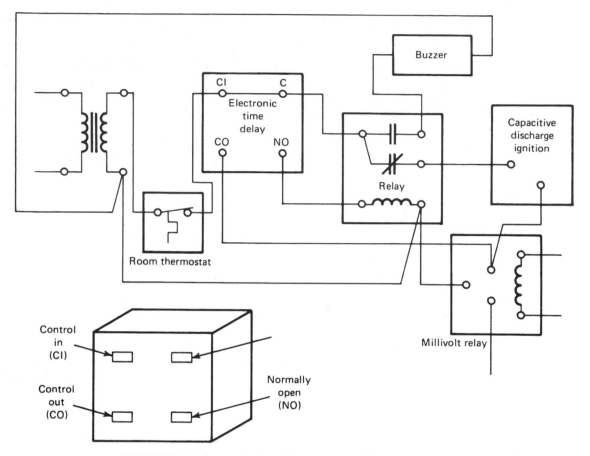

FIGURE 21-8 Electronic time delay circuit with external relay.

increasing pressure. This pressure controls a bellows-actuated switch. The switch, in turn, controls the application of power to a relay coil. The hydraulic flame sensor is shown in Figure 21-9. A part of the control circuit that includes the hydraulic flame sensor is shown in Figure 21-10. The switch controls the application of power to the relay, and the relay cuts off the capacitive discharge system. The relay also completes the circuit that opens the main burner valve.

Electronic Flame Sensor

The purpose of a flame sensor is to prove that the pilot is lit. The electronic flame sensor does this by using the pilot flame as the medium of electrical conduction. An electronic flame sensor is shown in Figure 21-11.

It is a well-known fact that a flame is a chemical reaction. In the chemical reaction of a gas pilot flame, positive electrical ions are developed. In a circuit, the positive ions are attracted to the conducting surface with the largest surface area. The movement of the positive ions to the conducting surface with the largest area provides a

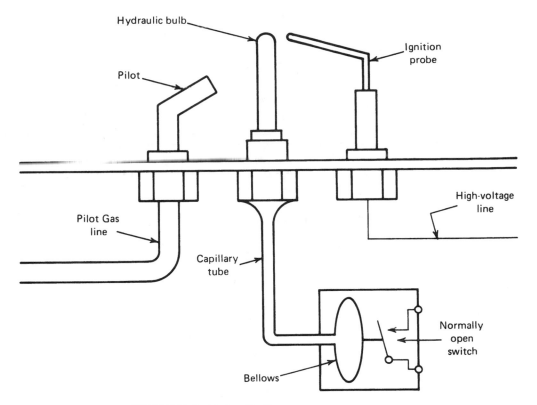

FIGURE 21-9 Hydraulic flame sensor.

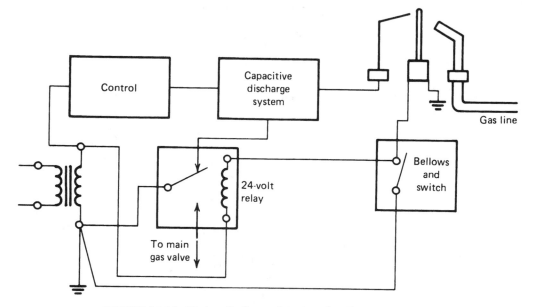

FIGURE 21-10 Hydraulic flame detector circuit.

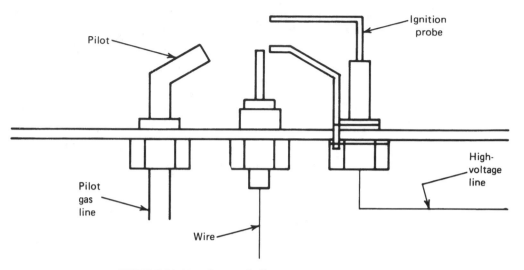

FIGURE 21-11   Electronic flame sensor.

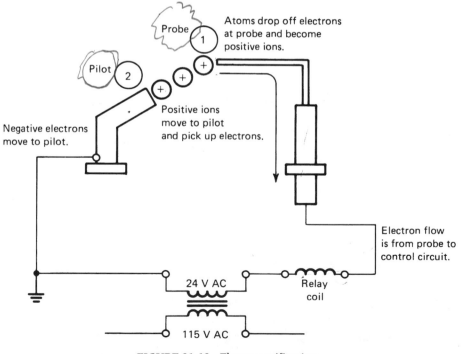

FIGURE 21-12   Flame rectification.

conduction path in one direction. The process is sometimes referred to as *flame rectification*. In the gas pilot system, the metal pilot burner is the metal surface with the largest area. The positive-ion movement is from the smaller probe to the pilot burner. The pilot burner is grounded. The circuit action is shown in Figure 21-12.

The positive ions continuously form in the flame by dropping off electrons at the probe (1). The positive ions move to the metal housing of the pilot light (2) where they pick up electrons, thereby becoming a neutral atom again. The process continues with electrons moving from ground to the pilot, and electrons then leaving the probe through the external circuit.

In the external circuit, the electron current is fed from the probe through the relay coil and through the low-voltage transformer secondary, as shown in Figure 21-13. The relay in Figure 21-13 provides the same function as the millivolt relay

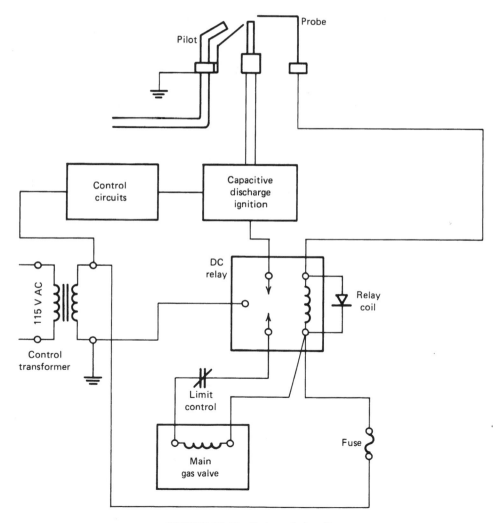

FIGURE 21-13   External circuit.

shown in Figures 21-6 and 21-8. The purpose is to control the shutdown of the capacitive discharge ignition and to complete the control circuit, thus opening the main burner's solenoid gas valve.

A safety feature added to this circuit is the diode placed across the relay coil. The diode is placed in the circuit because the normal electron flow resulting from flame rectification cannot pass through it. This situation is shown in Figure 21-14a. Remember that electrons can only flow through a diode in the direction toward the arrow head. The electron flow resulting from the flame rectification process cannot go through the diode; it must go through the relay coil. If the probe should be bent, making contact with ground (short circuit), there would be a current path during both alternations of the ac input. On every other alternation of the input, there would be a short path through the diode, as shown in Figure 21-14b. Sufficiently high current would flow, causing the

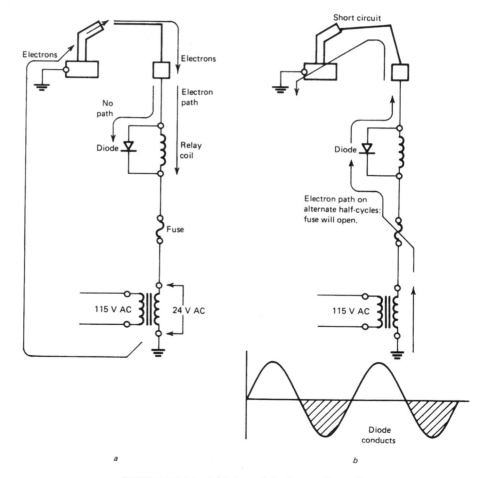

FIGURE 21-14 Addition of diode to relay coil.

fuse to open. All power would then be removed from the system, and it would be shut down.

New controls are rapidly being developed for the heating industry. Until such a time when the controls become standardized, it is necessary that the technician always refer to the manufacturer's information sheet concerning the particular unit being installed or serviced.

HOT-SURFACE IGNITION

"Hot-wire ignition"

Hot-surface ignition is a recent development in the gas-furnace control field. The ignitor is made from high-purity recrystallized silicon carbide. This material provides for both physical and thermal strength. The material also has stable electrical properties.

Electrical connection to the ignitor is made with nickel chrome lead wires. The wires are embodied in the silicon carbide and metallized in place. The lead wires are covered with high-temperature–resistant fiberglass insulation. The ignitors are available with wire leads or terminal connections as required by individual application. See Figure 21-15.

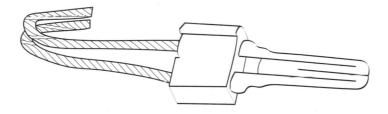

FIGURE 21-15   Hot surface ignitor.

The hot-surface ignitor is used along with a companion ignition module. Typical operation of the system is as follows.

When the thermostat calls for heat, the closed contacts within the thermostat feeds 24-V ac to the ignition module. If the model is of the prepurge type, there will be a delay before power is applied to the hot-surface ignitor. The combustion blower or other evacuation device will be turned on to ensure the removal of concentrated flammable gas from the system. After the delay, power is applied to the ignitor. When the ignitor is hot, the gas valve will open. Power will be applied to the ignitor for a fixed period of time, called the trial period.

When main burner ignition occurs, the flame is sensed by the ignitor or by a remote sensor. The ignitor is then turned off. If no flame is detected within the trial period, the gas valve's main valve is closed, shutting off the gas flow.

If ignition is not accomplished during the trial period, the thermostat must be set to a position (heat not called for) for a period of time to 3 seconds prior to establishing a new attempt to start.

SUMMARY

There are many types of control devices used in gas furnaces. The manufacturers' service data provide the best sources of maintenance information.

PRACTICAL EXPERIENCE

Required equipment   Gas furnaces, manufacturers' maintenance data.

Procedure

1. In the maintenance data, determine the system that controls the furnace.

2. Follow the circuit diagram while at the same time locating the operating and control elements of the furnace within the furnace.

REVIEW QUESTIONS

1. The thermocouple produces a gas output that controls the pilot flame.   T_____   F_____

2. The thermopilot gas valve feeds gas directly to the main burner.   T_____   F_____

3. The main burner gas valve is controlled by either the single thermocouple or a thermopile in the millivolt system.   T_____   F_____

4. The high-voltage capacitive discharge ignition operates whenever its main burner is turned on.   T_____   F_____

5. A plenum heat limit thermostat is not needed in a capacitive discharge ignition gas system.   T_____   F_____

6. All electronic time delays must be modified for use with gas heating systems.   T_____   F_____

7. A time delay prior to establishing ignition is used to exhaust flammable gases from the combustion chamber.   T_____   F_____

Conditioned Air Delivery

Objectives

Upon completion and review of this unit, you will understand that
- Conditioned air is air that is at the proper temperature and humidity.
- Conditioned air is cleansed of pollutants such as dust, soot, smoke, and pollen.
- The air must be delivered to the spaces requiring conditioned air.
- The delivery system includes sensors, dampers, and motors.

A very important part of an air-conditioning technician's responsibility is the delivery of conditioned air to the space to be heated or cooled. To a great extent, the overall efficiency of the system is dependent on how well an air delivery system operates. It is of little benefit to have an evaporator system operating at maximum efficiency if the cooled air does not reach the space to be air conditioned; more important, there is no reason to have the evaporator system operating at all if the outside air temperature is low enough to cool the space to be air conditioned. The air delivery system is the final phase of the complete heating and cooling system.

Figure 22-1 is the diagram of a complete conditioned air delivery system for an office or apartment building. The overall unit could be located in a penthouse or in a basement. The system would be approximately the same for both.

AIR-DELIVERY-SYSTEM COMPONENTS

The following paragraphs briefly describe the components of the air delivery system.

Air Handler

The air handler consists of the motor and fan system and associated controls. The air handler must provide the volume and air speed required by the system size.

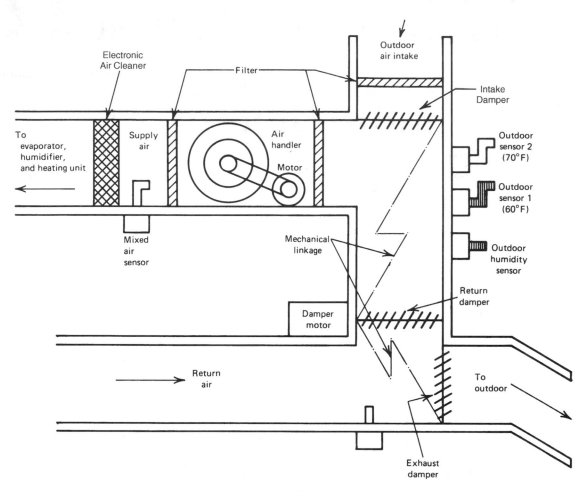

FIGURE 22-1 Conditioned-air delivery system.

Mixed Air Sensor

The mixed air sensor is a control whose operation is based on the temperature of the air being supplied to the conditioner. The mixed air is a combination of return air and outside air. This device, a thermostat, is usually set for about 60°F (15.55°C).

Outdoor-Ambient-Temperature Sensor 1

Outdoor-ambient-temperature sensor 1 is a control whose operation is based on whether the "free" air is cold enough to be used for cooling. This device is usually set for about 60°F.

Outdoor-Ambient-Temperature Sensor 2

Outdoor-ambient-temperature sensor 2 limits the amount of outside air allowed into the system. When the temperature of the outdoor air reaches the preset limit, the outdoor dampers close to the minimum position, reducing the amount of warm outdoor air allowed into the system. This device is usually set for about 70°F.

Humidity Sensor

The humidity sensor controls the humidifier. The humidity sensor determines the amount of moisture in the return air. If the moisture level is below a preset limit, the humidifier will be turned on.

Outdoor Humidity Sensor

The outdoor humidity sensor senses the condition of the outdoor air and determines whether the compressor will be turned on when outdoor air is used for cooling.

Humidifier

The humidifier ensures that water will be introduced into the air flow as required. The humidifier is used for heating.

Exhaust Damper

The exhaust damper controls the amount of return air that is discharged to the outdoor through a duct, as shown in Figure 22-2.

Return Damper

The return damper controls the amount of air from the conditioned area that is going to be recirculated. The return damper is mechanically coupled with the exhaust damper. When the return damper closes, the exhaust damper opens.

Outdoor Damper

The outdoor damper controls the amount of outdoor air that is allowed to enter the system. A fixed amount of fresh air is always introduced into the system. The outdoor damper is mechanically linked to the return damper. When the return damper closes, the outdoor damper opens. When the system is off, the outdoor damper closes.

Damper Motor

The damper motor is designed specifically for damper system use. The standard motor has three wire inputs. The motor circuit is shown in Figure 22-3. When

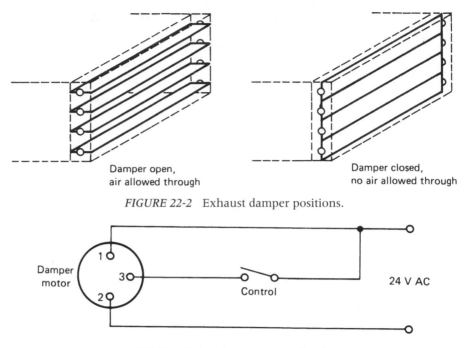

FIGURE 22-2   Exhaust damper positions.

FIGURE 22-3   Damper motor circuit.

power (24 volts) is applied between terminals 1 and 2, the motor will rotate to a limit in approximately 45 seconds. When terminal 3 is connected through the control circuit, the motor will rotate to the opposite limit in 45 seconds. When terminals 1 and 2 are active alone, the outdoor damper and exhaust damper will close and the return damper will be fully open. When terminal 3 is connected through the control circuit, the return damper will close and the outdoor and exhaust dampers will open.

MODULATING DAMPER ACTION

A modulating damper is in almost continuous motion, opening and closing in order to maintain the required mixed air temperature. The mixed air sensor controls the damper action as the air temperature at the sensor changes with changes in the damper position. The outdoor air damper, return air damper, and exhaust air damper will often be in continuous motion when the system is in the cooling mode of operation. The damper's actions are controlled by temperature measurements made with fixed temperature controls. Fixed temperature controls are shown in Figure 22-4.

CONTROL CIRCUITS

The system of air delivery shown in Figure 22-1 is an example of an energy-saving system. Whenever the outside air is cool enough, it will be used to cool or help cool the inside of the building. The operation of the system is controlled by

Thermostat symbols

Mixed air thermostat contacts open
when temperature falls below 60° F.

Outdoor air monitor contacts closed
when temperature falls below 70° F.

Mixed air thermostat contacts open
when temperature falls below 60°F.

>60°F

<60°F

Capillary
tube

Temperature-sensing
bulb

FIGURE 22-4   Fixed temperature controls—factory set.

temperature and humidity sensors located inside the air ducts and by sensors that respond to outdoor air temperature and humidity.

The complete circuit for this damper motor system is shown in Figure 22-5. As with other complete circuit diagrams, the energy-saving system circuit diagram appears complex. The circuit is easily understood when the individual units of operation are investigated.

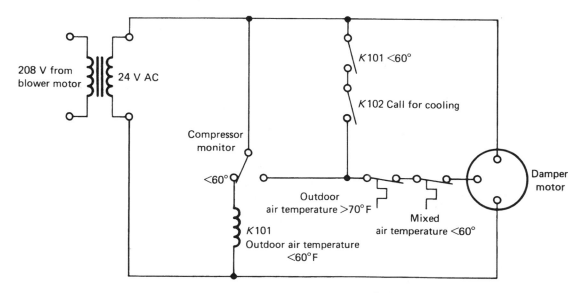

208 V from
blower motor

24 V AC

K101 <60°

K102 Call for cooling

Compressor
monitor

<60°

Outdoor
air temperature >70°F

Mixed
air temperature <60°

Damper
motor

K101
Outdoor air temperature
<60°F

FIGURE 22-5   Damper motor control.

DAMPER-MOTOR CONTROL

There are four basic weather conditions during which the damper control system is important:

1. Outdoor air temperature below 60°F (15.55°C) with low humidity.
2. Outdoor air temperature below 60°F with high humidity.
3. Outdoor air temperature between 60°F (15.55°C) and 70°F (21.11°C).
4. Outdoor air temperature above 70°F.

Outdoor Air Temperature Below 60°F with Low Humidity

The power source for the damper system is the step-down transformer that receives its primary voltage from the same source as the indoor blower motor. The connection of the transformer primary is shown in Figure 22-6. Whenever the indoor blower relay is energized, the damper low-voltage transformer primary will be connected across the 208-volt line. When an energy-saving damper system is used, the indoor blower motor operates continuously whenever the main disconnect switch is closed. This is equivalent to having the thermostat fan

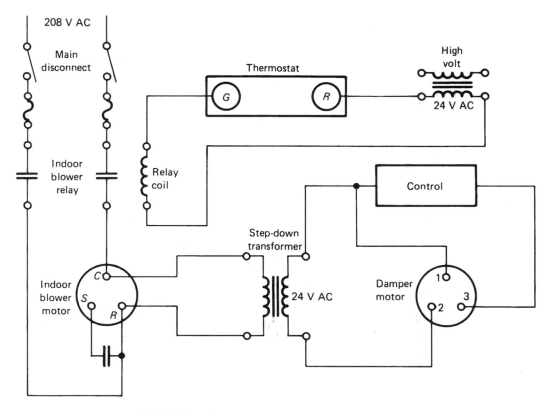

FIGURE 22-6   Damper motor power input.

switch in the on position. The 24 volts ac is then available at the secondary to operate the damper motor. With the 24 volts ac connected to terminals 1 and 2 of the damper motor, the motor rotates to the start limit. In this position, the return damper is fully open. The exhaust damper is closed, and the outdoor damper is opened the minimum amount, providing a limited amount of fresh air to the system.

An outdoor air temperature below 60°F (15.55°C) is low enough to meet the cooling needs of the building without the operation of the compressor. The damper control system must keep the compressor from operating and must open the outdoor damper, close the return damper, and open the exhaust damper.

The control of the compressor is provided by relay $K103$ (see Figure 22-7). The normally open contacts 1 and 3 of relay $K103$ will be open if the outside air temperature is below 60°F (15.55°C). These contacts are in series with the $Y1$ output of the thermostat. A complete circuit will not exist to the compressor

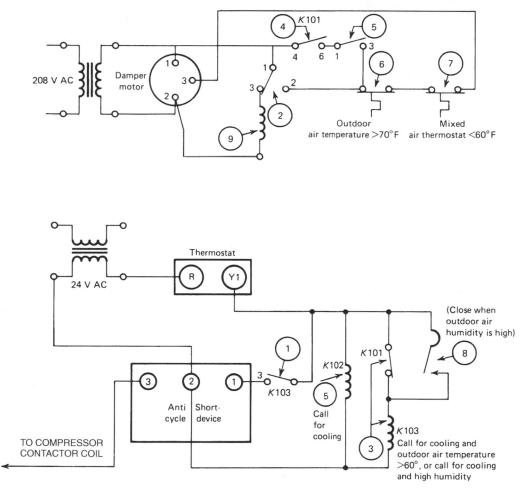

FIGURE 22-7 Compressor and damper control.

contactor coil. With the contacts of *K*103 open, the compressor contactor cannot be energized. The contacts of relay *K*103 are at number 1 in Figure 22-7.

Relay *K*103 is controlled by the outdoor air thermostat, numbered 2 in Figure 22-7. When the outdoor temperature is below 60°F (15.55°C), the 1 and 3 contacts of the thermostat switch will be closed, providing 24 volts ac across relay *K*101. Relay *K*101 will energize and open the normally closed contacts 1 and 2, which are in series with the coil of relay *K*103. This is shown as 3 in Figure 22-7. Relay *K*103 will stay de-energized, since relay *K*101 is energized and the outdoor air has low humidity.

The resulting situation is that the thermostat has a call for a cooling condition. The compressor will not operate the relay, since *K*103 contacts are open. The thermostat output at *G* calls for indoor blower operation, which provides a 208-volt input to the damper supply transformer (see Figure 22-6). The damper motor is in the normal position, the return damper is open, and both exhaust damper and outdoor damper are closed. In order to obtain cooling from outside air, the outdoor damper must be opened, the return damper closed, and the exhaust damper opened. This can be accomplished by providing 24 volts ac to terminal 3 of the damper motor. Relay *K*101 shown at 2 in Figure 22-7 is energized, since the outdoor air temperature is below 60°F. A set of normally open contacts 4 and 6 of relay *K*101 is shown (at 4 in Figure 22-7) in series with the normally open contacts 1 and 3 of relay *K*102 (at 5 in Figure 22-7). Relay *K*102 is energized whenever there is a call for cooling.

The circuit to terminal 3 of the damper motor continues through the contacts of the thermostat outdoor air temperature greater than 70°F (21.11°C) and the contacts of the mixed air thermostat. This part of the damper circuit is shown in Figure 22-8. The damper motor will rotate and open the outdoor damper. The return damper will close and the exhaust damper will open.

When the outdoor damper is opened, cool outdoor air (temperature below 65°F—18.33°C) is introduced into the cooling system. The return damper is closed, blocking the warmer indoor air from recirculating, and the exhaust damper opens to release the warm indoor air to the outdoors. This system is shown in Figure 22-9.

After sufficient outdoor air has entered the system, the cooling effect will reduce the mixed air temperature to below 60°F (15.55°C). The thermostat will open, and the circuit to terminal 3 of the damper motor will be incomplete. Since

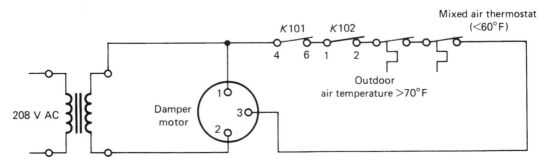

FIGURE 22-8   Damper connections for use of outdoor air.

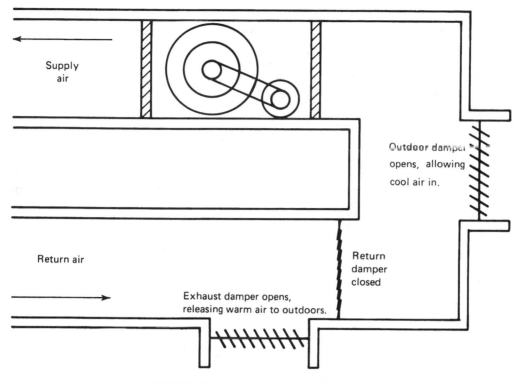

FIGURE 22-9 Damper positions for use of outdoor air.

power is still being applied to terminals 1 and 2 of the damper motor, the motor will rotate to close the outdoor damper. At the same time, the return dampers will be opening, and the exhaust dampers will be closing. Warmer return air will be mixed with outdoor air. The outdoor damper will continue to close, and the return damper will continue to open until the air reaching the mixed air thermostat (as shown in Figure 22-8) reaches a temperature above 60°F. The thermostat will then close again. The damper will again have a power connection at terminal 3 and will start to rotate in a direction to open the outdoor damper, allowing in more cool outdoor air.

The damper system will modulate, opening and closing, thus keeping the temperature at the mixed air thermostat varying around 60°F (15.55°C). The inside of the building will be cooled solely by the cooler outdoor air. The compressor will not operate and energy will be saved.

High Humidity

The use of outdoor air for cooling when the humidity is high is practical, but some water must be removed from the air if the system is to provide comfortable air—that is, the outdoor air must be treated to reduce the humidity. To accomplish this, the compressor system is allowed to operate even though it is not needed for cooling. When the outdoor air moves across the much cooler evaporator coils, the air will be further chilled, causing water to condense from it. The cooled air is then fed inside the building, where it mixes with warmer air. The combined air is cool and low in humidity.

The circuit that provides system control when the outdoor air temperature is below 60°F (15.55°C) and humidity is high requires only the addition of the humidity sensor contacts (numbered 8 in Figure 22-7). They are normally open contacts, which close when the humidity exceeds a preset limit.

When there is high humidity, relay $K103$ will close every time there is an output from the $Y1$ terminal of the thermostat (that is, a call for cooling). The circuit to the coil of relay $K103$ is through the contacts of the humidity sensor. The damper system will operate as before, with the damper motor rotating to open and close the damper. The mixed air temperature sensor continues to control the direction of motor rotation. The dampers modulate, keeping the air temperature at the mixed air temperature sensor around 60°F.

NOTE: *The symbol for* greater than *is* >. *The symbol for* less than *is* <.

Outdoor Air Temperature Between 60°F (15.55°C) and 70°F (21.11°C)

If the outdoor air temperature is between 60°F and 70°F, the air is still useful for cooling purposes but is not sufficiently cool to satisfy all cooling needs—this is assuming that the return air will have a temperature above 70°F. Such a situation is normal when cooling is called for in most systems.

In order to make use of the available cooling from outdoor air, the outdoor damper should be fully open, the return damper fully closed, and the exhaust damper fully open. With the dampers in the indicated positions, the maximum amount of outdoor air (below 70°F) will be admitted to the system. The warmer return air will be blocked by the closed return damper. The warm return air will be discharged to the outdoors, through the exhaust damper. The circuit connection for accomplishing the required damper movements is shown in Figure 22-10.

Since the outdoor air temperature is between 60°F and 70°F, the outdoor air > 70°F thermostat will not activate. The contacts of the outdoor air temperature below 60°F thermostat will be in the position shown in Figure 22-10. The mixed air temperature is above 60°F. The mixed air temperature below 60°F thermostat will not activate. A complete circuit exists to terminal 3 of the damper motor. The motor will rotate to the position where the outdoor damper is open, the return damper closed, and the exhaust damper open. Any further action of the air-conditioning system should not cause a change in the dampers.

Outdoor Air Temperature Above 70°F

Whenever the outdoor air temperature is above 70°F, it is considered (in this energy-saver system) to be unsuitable for cooling purposes. The desired damper conditions are: outdoor damper closed, return damper open, and exhaust damper closed. These conditions are met by the action of the contacts associated with the

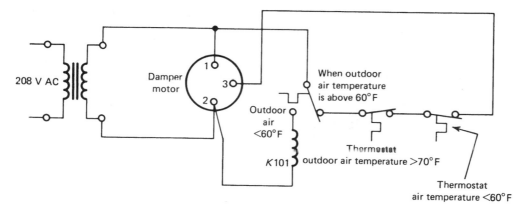

FIGURE 22-10 Outdoor air temperature between 60°F and 70°F.

outdoor air temperature greater than 70°F. These contacts are shown closed in Figure 22-10. When the outdoor air temperature is above 70°F, the contacts will open. This action will interrupt the circuit to 3 of the damper motor. The motor will rotate, closing the outdoor damper. The return damper will open, and the exhaust damper will close.

The compressor will operate whenever there is a call for cooling. Relay $K101$ is de-energized, and its 1 and 2 contacts are closed. The circuit to the coil of relay $K103$ is complete. It will energize whenever the thermostat calls for cooling. The 1 and 3 contacts of $K103$ are closed, completing the circuit to the compressor contactor coil.

Electronic Control of Damper Systems

A number of electronically controlled damper systems are presently being developed. The use of electronic control allows more sensitive detecting of conditions.

With a thermostat that operates a set of switch contacts mechanically, the differential of turn on and turn off may be 3 or 4°F. This differential may be reduced by using an electronic sensor and amplifier system.

There are control units presently available that sense return air temperature and humidity and compare it with outdoor air temperature and humidity. The selection of outdoor air or return air is based on which contains less *total* heat. *(enthalpy control)*

The introduction of new electronic control systems is proceeding at a rapid pace. Standardization of these control systems may take place in the future. For the present, however, each system must be studied individually.

Filters

Filters are used to keep foreign particles that might clog up the heating and cooling elements out of the system.

The standard filters used with most conditioned air delivery systems remove large particles such as hair, lint, and dirt from the recirculated air. Some of the smaller dust particles are also removed by standard filters. Many particles do, however, get through the standard filter and are recirculated through the system.

Electronic Air Cleaners

Electronic air cleaners remove from the redistributed-air fine particles such as dust, soot, smoke, and pollen that have made their way through the standard filter.

Electronic Air Cleaner Operation   Unit I of this book covered the concept: Like charges repel each other while unlike charges attract.

Electronic air cleaners function using this principle. The air to be cleaned is passed through an ionizing section where most particles pick up a positive charge. The charging section (ionizer) usually consists of positive-charged wires. Electrons are accelerated towards the positive-charged wires. Some of the fast-moving electrons collide with other electrons in the air molecules, knocking them loose from the molecule. This action leaves many positively charged air molecules. See Figure 22-11. The positively charged air molecules become attached to unwanted dirt particles. The *air* and *dirt* molecules are fed through an area of an electric field where the positively charged particles

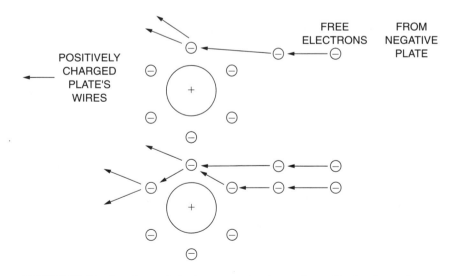

FIGURE 22-11   Accelerated electrons moving towards positively charged wires knock off electrons, leaving positively charged air molecules.

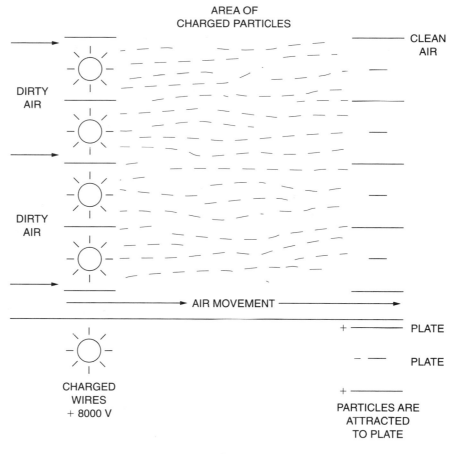

FIGURE 22-12 Internal operation of air cleaner.

move through the field to a negatively charged plate. (See Figure 22-12). In Figure 22-13, the high-voltage section of a typical electronic air cleaner may be observed.

The transformer is ferroresonat. The secondary resonate winding provides for nearly constant output voltage and also limits output current. Typical output voltage is 4000 V peak or 8000 V peak to peak. The rectifier circuit produces 8000 V dc between the positive and negative terminals. The output current limit is about 5 mA.

As with most devices, periodic maintenance is required. To remain effective, electronic air cleaners must be cleaned of dirt that has been collected. Many of the parts may be washed. It is important that the manufacturing maintenance procedures be followed. Repair of malfunctioning units is best accomplished following the manufacturer's set of procedures.

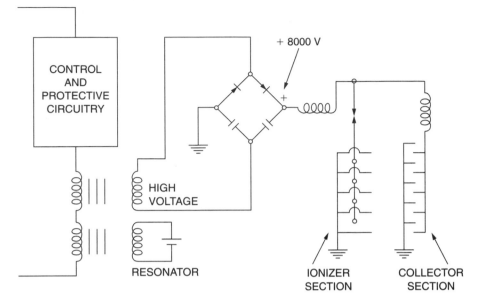

FIGURE 22-13 Electronic air cleaner, high voltage.

SUMMARY

1. The conditioned air delivery is the final product of any air-conditioning system.

2. It is important to understand the complete air-delivery system.

3. There is little value to having a high-efficiency heating or cooling system if the conditional air does not reach the proper location.

REVIEW QUESTIONS

1. In most air conditioning systems, the dampers operate independently of each other.   T_____   F_____

2. The exhaust damper provides a path for the mixed air.   T_____   F_____

3. The outdoor damper is closed tight whenever the outdoor air temperature is above 65°F.   T_____   F_____

4. If the outdoor air temperature is below 60°F, the dampers are expected to modulate.   T_____   F_____

5. When outdoor air alone is used for cooling, the outdoor damper should start to close when the mixed air temperature goes below 60°F.   T_____   F_____

6. The return damper never closes.   T_____   F_____

7. It is normal to have the outdoor damper open slightly whenever the system is in operation.   T_____   F_____

Appendix A

Electrical Symbols Common to Air-Conditioning Systems

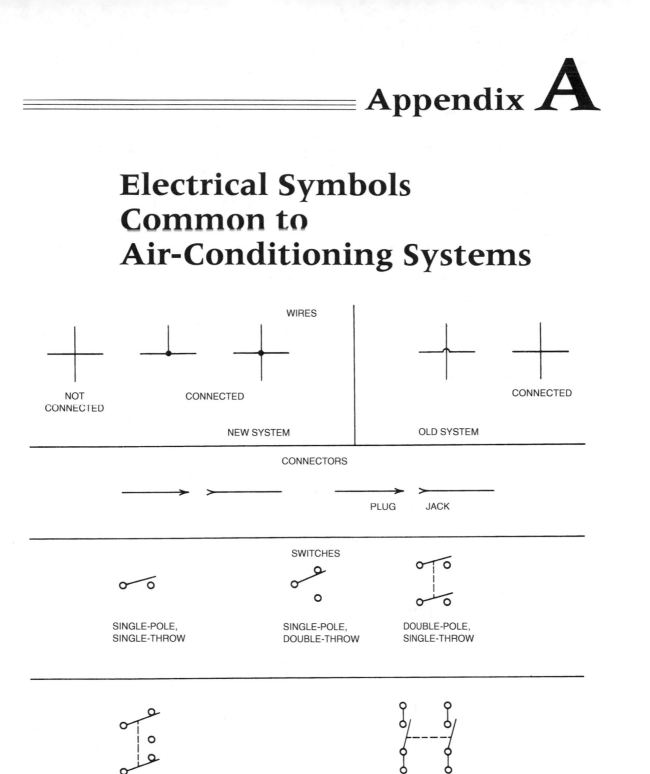

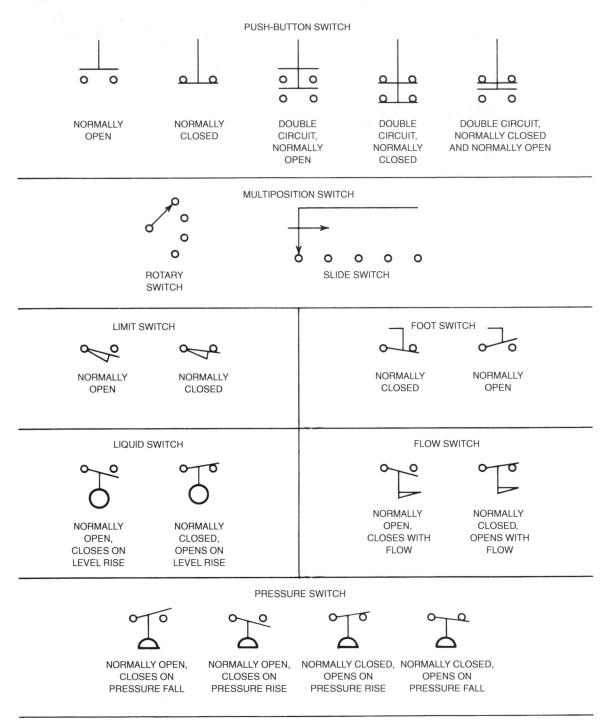

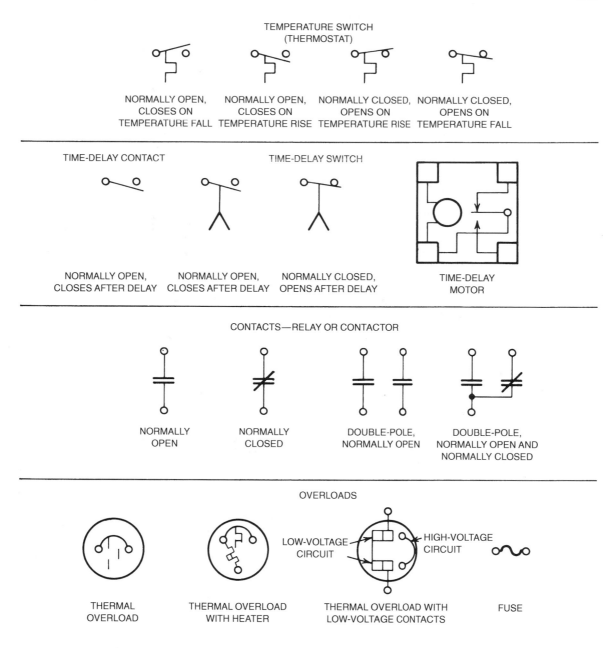

TEMPERATURE SWITCH
(THERMOSTAT)

NORMALLY OPEN,
CLOSES ON
TEMPERATURE FALL

NORMALLY OPEN,
CLOSES ON
TEMPERATURE RISE

NORMALLY CLOSED,
OPENS ON
TEMPERATURE RISE

NORMALLY CLOSED,
OPENS ON
TEMPERATURE FALL

TIME-DELAY CONTACT

TIME-DELAY SWITCH

NORMALLY OPEN,
CLOSES AFTER DELAY

NORMALLY OPEN,
CLOSES AFTER DELAY

NORMALLY CLOSED,
OPENS AFTER DELAY

TIME-DELAY
MOTOR

CONTACTS—RELAY OR CONTACTOR

NORMALLY
OPEN

NORMALLY
CLOSED

DOUBLE-POLE,
NORMALLY OPEN

DOUBLE-POLE,
NORMALLY OPEN AND
NORMALLY CLOSED

OVERLOADS

LOW-VOLTAGE
CIRCUIT

HIGH-VOLTAGE
CIRCUIT

THERMAL
OVERLOAD

THERMAL OVERLOAD
WITH HEATER

THERMAL OVERLOAD WITH
LOW-VOLTAGE CONTACTS

FUSE

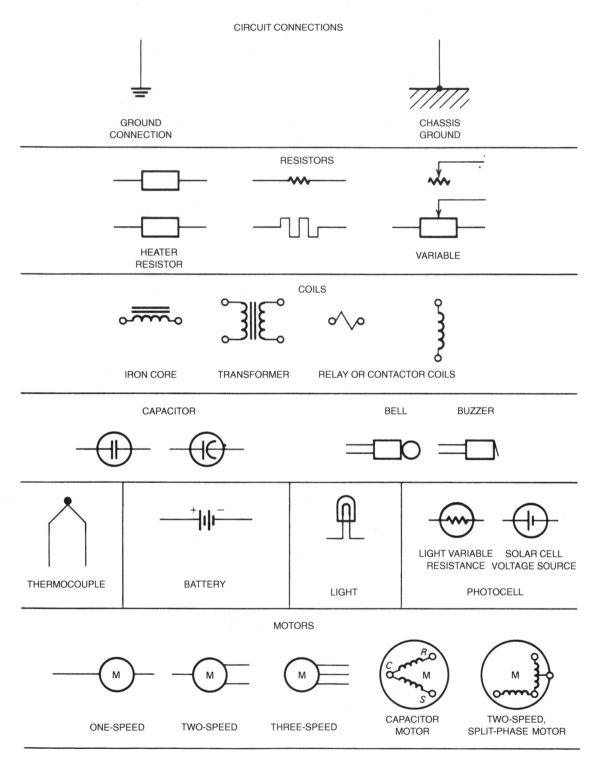

CIRCUIT CONNECTIONS

GROUND
CONNECTION

CHASSIS
GROUND

RESISTORS

HEATER
RESISTOR

VARIABLE

COILS

IRON CORE   TRANSFORMER   RELAY OR CONTACTOR COILS

CAPACITOR   BELL   BUZZER

THERMOCOUPLE   BATTERY   LIGHT   LIGHT VARIABLE   SOLAR CELL
                                                          RESISTANCE   VOLTAGE SOURCE

                                                                    PHOTOCELL

MOTORS

ONE-SPEED   TWO-SPEED   THREE-SPEED   CAPACITOR   TWO-SPEED,
                                                          MOTOR   SPLIT-PHASE MOTOR

THREE-PHASE MOTOR CONNECTIONS

THREE-PHASE
WYE

THREE-PHASE
DELTA

SEMICONDUCTORS

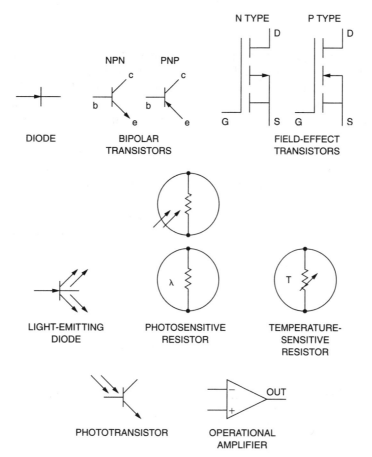

DIODE

NPN   PNP

BIPOLAR
TRANSISTORS

N TYPE   P TYPE

FIELD-EFFECT
TRANSISTORS

LIGHT-EMITTING
DIODE

PHOTOSENSITIVE
RESISTOR

TEMPERATURE-
SENSITIVE
RESISTOR

PHOTOTRANSISTOR

OPERATIONAL
AMPLIFIER

Appendix B

Powers of Ten

The use of the powers of ten provides for mathematical operations using very large or very small numbers. In electronics most calculations are held to three significant digits, that is, in the number 3,262,137,812, only the 326 and the position of the decimal would be important in most situations. Electronic components and measurements are usually only accurate to three significant digits.

Considering three-significant-digit use in electronics makes the powers of ten very important. The number 3,262,137,812 becomes 3.26×10^9. The decimal is simply moved to the point, providing a single-digit whole number. The number of places the decimal has been moved provides the power of ten. For numbers less than 1, say, 0.0000000192, this becomes 1.92×10^{-8}.

Numbers with a power of ten may be added if the numbers have the same power of ten.

$$
\begin{array}{r}
2.13 \times 10^4 \\
+ \ 1.33 \times 10^4 \\
\hline
3.46 \times 10^4
\end{array}
$$

If the powers of ten are different, they must be made the same before adding.

EXAMPLE 1

Add (2.14×10^6) plus (1.21×10^5).

Solution: Change 1.21×10^5 to 0.121×10^6.

$$
\begin{array}{r}
0.121 \times 10^6 \\
+ \ 2.14 \ \ \times 10^6 \\
\hline
2.26 \ \ \times 10^6
\end{array}
$$

The same conditions exist in subtraction.

EXAMPLE 2 _____

Subtract 8.37×10^5 from 1.28×10^6.

Solution: Change 1.28×10^6 to 12.80×10^5.

$$
\begin{array}{r}
12.80 \times 10^5 \\
-\ 8.37 \times 10^5 \\
\hline
4.43 \times 10^5
\end{array}
$$

When powers of ten are used in multiplication, the numbers are multiplied and the powers of ten are added.

EXAMPLE 3 _____

Multiply (3.11×10^4) times (2.06×10^2)

Solution:

$$
\begin{array}{r}
3.11 \quad \times 10^4 \\
\times\ 2.06 \quad \times 10^2 \\
\hline
6.4066 \times 10^6 \quad \text{or} \quad 6.41 \times 10^6
\end{array}
$$

EXAMPLE 4 _____

Multiply (2.17×10^{-4}) times (4.23×10^3).

Solution:

$$
\begin{array}{r}
2.17 \quad \times 10^{-4} \\
\times\ 4.23 \quad \times 10^3 \\
\hline
9.1791 \times 10^{-1} \quad \text{or} \quad 9.18 \times 10^{-1}
\end{array}
$$

In division, the number is divided by the divisor and the powers of ten are subtracted.

EXAMPLE 5 _____

Divide (6.11×10^6) by (2.07×10^3).

Solution:

$(6.11 \times 10^6) \div (2.07 \times 10^3) = 2.95 \times 10^3 \quad \text{or} \quad 2950$

EXAMPLE 6 _____

Divide $(3.78 \times 10^5) \div (2.16 \times 10^{-3})$.

Solution: $(3.78 \times 10^5) \div (2.16 \times 10^{-3}) = 1.75 \times 10^8$

POWERS OF TEN

10^{10}	10,000,000,000
10^9 Giga	1,000,000,000
10^8	100,000,000
10^7	10,000,000
10^6 Mega	1,000,000
10^5	100,000
10^4	10,000
10^3 Kilo	1,000
10^2	100
10^1	10
10^0	1
0	0
10^{-1}	0.1
10^{-2}	0.01
10^{-3} Milli	0.001
10^{-4}	0.0001
10^{-5}	0.00001
10^{-6} Micro	0.000001
10^{-7}	0.0000001
10^{-8}	0.00000001
10^{-9} Nano	0.000000001
10^{-10}	0.0000000001

Glossary

A   Abbreviation for ampere.

ac   Abbreviation for alternating current. Also, sometimes used as an abbreviation for air conditioning.

alternating current (ac)   An electric current that alternates. Its flow is considered to change direction with each half cycle.

alternator   A device that produces alternating current by rotating magnets past stationary coils of wire.

ammeter   An instrument used to measure the rate of current flow in a circuit.

amp   Abbreviation for ampere.

ampacity   Current-carrying capacity, expressed in amperes.

amperage   The total amount of current flowing in a circuit.

ampere   The unit of measure for the rate of flow of electrons. This value is obtained by dividing circuit voltage by its resistance in ohms.

ampere-hour   A unit of measure for battery capacity. This value is obtained by multiplying the current, in amperes, by the time, in hours, during which the current is delivered.

amp-hour   An abbreviation for ampere-hour.

amplifier   An electronic device used to control the amplitude of a signal, either voltage or current.

arc   A flow of ions through the air between two electrodes or contact points. Material, usually metal, is transferred from one point to another.

armature   That part of a generator, starter, or motor that rotates inside the unit.

atom   The smallest particle of matter that has the characteristic chemical properties of an element.

back emf (bemf)   An induced voltage in a circuit or component that is in direct opposition to the applied voltage.

battery   An electrochemical device for storing mechanical energy in chemical form so that it can be released as electricity when needed.

bimetallic   Relating to two dissimilar metals, usually invar and copper, that have been fused together. Changes in temperature cause a bimetallic strip to change shape (bend).

brush   A conductor arranged so as to make contact with a rotating surface, such as with the armature of a motor or the slip ring of an alternator.

brush holder   A device used in motors and generators to hold the brushes in position on the commutator.

cable   Insulated wire, usually stranded, used for conducting electrical current.

capacitor   A device used to temporarily store an electrical charge.

capacity   The quantity of available electricity that can be delivered from a system or device, such as ampere-hours from a battery.

carbon pile   A series of carbon discs held loosely together by an insulated holder, generally found in a battery load tester. The tighter the discs are pressed together, the less the resistance to current flow.

cell   A combination of two dissimilar metals in an electrolyte. Two or more cells make a battery.

Celsius (C)   A temperature scale used in the metric system. The freezing point of water at sea-level atmospheric pressure is 0°C, and the boiling point is 100°C.

Centigrade   A term often used for the Celsius scale of temperature measurement.

centrifugal force   That force that tends to impel an object or parts of an object outward from a center of rotation.

centrifugal short   A shorted winding of an armature or rotor that only occurs when centrifugal pressure builds up while the unit is rotating.

charging rate   The amount of amperage flowing from the generator or alternator to charge a battery.

circuit   The path of electron flow from the source through the wiring, components, and connections and back to the source.

circuit breaker   An electrical overload protective device for interrupting a circuit when the current flow exceeds safe limits for a particular circuit.

circuit ground   The portion of a circuit used as a common potential point.

clockwise (cw)   A rotation in the same direction as the hands of an analog clock.

closed   Not open. Generally refers to a set of contacts that are connected in a switch or relay.

closed circuit   A completed circuit. A circuit in which there is no interruption of current flow.

color code   The system of color used on wires and some connectors of electrical circuits to aid and facilitate the tracing and troubleshooting of electrical system and subsystem problems. Or the system of colored bands on electrical components, used to indicate the component's electrical value.

common   Another term used for *ground.*

commutator   The part of a generator or motor armature that consists of copper bars. Brushes ride on the commutator to collect or supply current to the armature windings

condenser   A term often used, though not proper, for *capacitor.* That part of the air conditioning system that dissipates heat collected in the evaporator.

conductor   A path for electrical current. Any material that allows electrons to flow.

continuity   A completed circuit or device that has no electrical interruption.

counterclockwise (ccw)   Rotation in a direction opposed to the direction of the hands of an analog clock.

current   The flow of electrons in a circuit.

current limit relay   A relay that protects a circuit from overload by opening its contacts and breaking the circuit when the current flow reaches a predetermined high.

cycle   A complete event, from start to finish.

dc   Abbreviation for direct current.

de-energized   A term given an electrical control device, such as a relay, when no power is applied to the coil.

delay relay   See *time-delay relay.*

density   Quantity per unit of volume; compactness of matter.

dielectric   A nonconductor of electricity, such as an insulator.

diode   A device that permits current to flow in one direction only. A kind of electrical one-way check valve.

diode, zener   See *zener diode.*

direct current (dc)   An electrical current, as produced by a generator or battery, which flows in one direction only.

direction of electron flow   The flow of electrons, in a circuit, from negative to positive potentials.

discharge   To remove energy from a capacitor.

distilled water   Water that is free of impurities by distillation—first heated to a steam vapor and then cooled back to liquid.

draw   The amount of current required to operate an electrical device is often referred to as current draw.

drop in voltage   See *voltage drop*.

electrical horsepower   A measure of electrical power. One horsepower is equal to 746 watts.

electricity   A form of energy produced by chemical change, induction, heat, light, or friction.

electrode   One of the conductors of a cell.

electrolyte   A liquid conduction medium in which the flow of current is accompanied by a movement of material in the form of ions.

electromagnet   A coil of wire with a soft iron core, capable of being magnetized only as long as current is applied to the coil.

electromotive force   That voltage (force or pressure) that causes electron movement in an electrical circuit.

electron   The negatively charged, lightweight, movable particle of an atom.

electronics   The branch of physics that studies the behavior of and applies the effects of the flow of electrons.

element   A substance that cannot be separated into two or more substances.

emf   Abbreviation for *electromotive force*.

energized   A term given an electrical control device, such as a relay, when power is applied to its coil.

energy   Capacity for performing work. Electrical energy is measured in watt-hours.

evaporator   That part of an air-conditioning system that picks up unwanted heat.

Fahrenheit (F)   A temperature scale used in the English system. The freezing point of water at sea-level atmospheric pressure is 32°F, and the boiling point is 212°F.

field   A term referring to an area under the influence of an electrically charged object. Also, see *magnetic field*.

field coil   That part of a motor or generator that is mounted inside and to the frame. It is positioned around the armature.

field, magnetic   See *magnetic field*.

filament   The resistive element of a bulb that glows to produce light when current is forced through it.

flux   A substance used in soldering, brazing, and welding that promotes fusibility. Also, see *magnetic flux*.

flux density   The quantity of force per unit area. See *magnetic flux*.

flux, magnetic   See *magnetic flux*.

foot   A unit of measure. There are 12 inches to a foot.

foot pounds   A measure of torque or turning force; measured by a prony brake.

force   Energy acting to produce power or acceleration.

frequency   In alternating current, the number of cycles per second. Frequency is measured in hertz.

fuse   An electrical overload protective device for interrupting a circuit when the current flow exceeds safe limits for a particular circuit. When overloaded, the fuse will melt to open the circuit.

gauge   A measuring device, such as for wire, used to check size. Or a mechanical dial-type measuring device used to check pressure.

generator   A device that is used to convert mechanical energy into electrical energy.

germanium   A semiconductor material.

ground   The common in an electric circuit; in power distribution systems, the earth—or electrical connections to the earth.

ground circuit   The wiring or body metal that makes up the ground of an electrical circuit.

grounded circuit   A circuit that is intentionally or unintentionally shorted to ground.

growler   An electrical device used to check an armature to determine whether its windings are open or shorted.

hall effect   The change in conduction of a special semiconductor device when the device is in the influence of a magnetic field.

heat sink   A metal plate or bracket used to hold a heat-producing component, such as a diode, for the purpose of dissipating heat from the component.

helix   A term used to describe a spiral and cylindrical shape, such as a wire that was formed around a round core.

hertz (Hz)   The unit of frequency. Standard alternating current frequency in the United States is 60 hertz.

horsepower   A unit of energy. One horsepower is required to lift 550 pounds (249.5 k) 1 foot (0.3 m) in 1 second.

hot circuit   That portion of a circuit not at ground potential, and which is electrically insulated, and is usually at a potential (voltage).

hot lead   A wire or conductor connected to a high side, not to the ground or neutral terminal.

hot wire sensor   A device that senses air flow. A hot wire senses air flow by changing resistance when cooled by air movement.

hydrogen   See *hydrogen gas.*

hydrogen gas   The lightest and most explosive of all gases. Hydrogen (H) gas is emitted from a battery during charging and mandates that certain safety precautions be observed.

hydrometer   A device used to measure the specific gravity of the electrolyte in a battery.

inch   An English standard measure. There are 12 inches in 1 foot. One inch (1 in.) is equal to 25.4 millimeters (25.4 mm) in the metric measure.

inductance   The property in an electric circuit that opposes a change in current.

induction   The current induced in a conductor when the conductor cuts across a magnetic field or the magnetic field moves across the conductor.

inertia   The tendency of a body at rest to remain at rest and a body in motion to remain in motion.

insulate   To bar the flow of electrical current by the use of a nonconducting material.

insulation   Nonconducting materials, such as the plastic covering on wire, used to prevent the passage of electrical current.

insulation tape   A plasticlike tape used to insulate bare or exposed wire.

insulator   A nonconducting material used to prevent the passage of electrical current.

ion   A positively charged atom.

junction   A connection of two or more wires.

junction block   An accessible block for the connection of two or more wires for the convenience of checking electrical circuits.

kilo-   A prefix used in the metric system to indicate 1000. For example, a kilogram is the same as 1000 grams.

kilometer   A metric measure used for distance. One mile, for example, is equal to 1.609 kilometers (1.609 km).

lamination   Placement of thin plates of metal next to each other to form the core of a coil or other electrical device.

lead   A malleable metal, bluish-gray in color. Its chemical symbol is Pb.

lead peroxide   A combination of lead and oxygen, chocolate brown in color. As found in the storage battery, the chemical symbol is PbO_2.

left-hand rule A method used to find the direction of electron flow or the direction of the lines of force around a current carrying wire. Also, see *right-hand rule*.

lines of force See *magnetic lines of force*.

load An electrical device connected into a circuit, and which draws current.

loadstone A natural magnetic iron ore which attracts and holds pieces of iron or steel, also spelled *lodestone*.

magnet A ferrous material, usually steel, with a north and south pole, having the properties of attracting other ferrous materials.

magnetic A material having the properties of being attracted to or magnetized by a magnet.

magnetic field The area around a magnet where magnetic lines of force occur.

magnetic flux See *magnetic field*.

magnetic lines of force See *magnetic field*.

magnetic poles The north (N) and south (S) ends of a magnet.

make A term used to indicate that an electrical control device, such as a switch or a set of points, is in the ON or CLOSED position.

matter That which occupies space and has weight. All things are made of matter.

mega- A prefix indicating 1 million (1,000,000).

meter The fundamental unit of length measure in the metric system. One meter is equal to 39.37 US inches. Also, an electrical measuring device, such as a voltmeter or an ammeter.

mica A nonmagnetic, nonconducting mineral material. Mica is often found as an insulator between riser bars of a commutator.

micro- A prefix indicating one-millionth ($\frac{1}{1,000,000}$).

mile An English measure of distance. One mile is equal to 5,280 feet (1,609 meters) or 1.609 km.

milli- A prefix indicating one-thousandth ($\frac{1}{1,000}$).

millimeter A metric measure for length equal to $\frac{1}{1,000}$ of a meter. One inch (1 in.), for example, is equal to 25.4 millimeters (25.4 mm).

motor A rotating electromagnetic device used to convert electrical energy into mechanical energy.

multiplex To use the same wire to carry more than one piece of information.

negative The electrical potential of a object that has gained some extra electrons, said to be negatively charged.

negative pole   That end of a magnet that would naturally orient itself with the magnetic south pole.

neutral   The quality of an object with an equal number of positive and negative particles.

neutron   A tightly bound collapsed combination of an electron and a proton.

nucleus   The central portion of an atom, which contains protons and neutrons.

off   See *open.*

ohm   A unit of electrical resistance that opposes the flow of current. The symbol of ohm is Ω.

ohmmeter   An electrical instrument used to measure resistance.

Ohm's law   The law of electricity determining the relationship between voltage, current, and resistance. Simply stated, voltage equals current times resistance ($E = I \times R$).

on   See *closed.*

open   A break, intentional or unintentional, in an electrical circuit. A circuit is intentionally opened with a switch or a set of contact points. No current will flow in an open circuit.

open circuit   See *open.*

open winding   A break that prevents the flow of current in the winding of a field, armature, or coil.

oscillate   To move or swing freely back and forth in a motion similar to that of a clock pendulum.

oscillograph   A device used to present graphically—through mechanical means—a display of electrical signals.

oscilloscope   A device used to present graphically—on a cathode-ray tube—a display of electrical signals.

parallel circuit   Two or more circuits, electrically side-by-side, whose positive and negative terminals are connected to their respective positive and negative sources.

permanent magnet   A magnet made of tempered steel, and which has the ability to hold its magnetism for long periods.

permeability   The ability of a material to conduct magnetic lines of force.

photocell   A semiconductor device that produces a voltage when exposed to light.

photo resistor   A semiconductor device that changes resistance when exposed to light.

piezoelectric The property of crystal material to produce a voltage when subjected to pressure.

polarity Refers to pole identification, such as N and S poles of a magnet or + and − poles of a battery.

polarize To give polarity, temporary or permanent, to a conductor or ferrous material.

pole Either end of a magnet or of an electrical potential. See *polarity*.

pole piece See *pole shoe*.

pole shoe That part of a rotating device, such as a motor, that is used to hold the field coils in their proper positions around the armature.

positive pole That part of a magnet that tends to orient itself with the magnetic north pole.

potential Electrical force, measured in volts. Potential may be positive or negative.

power The rate of doing work, usually expressed in horsepower.

pressure switch An electrical switch device that is actuated by pressure or a change in pressure.

primary winding The winding of a transformer that is connected to electric power.

prony brake A device used to measure foot pounds of torque.

proton A positively charged particle found in the nucleus of all atoms.

RAM Random access memory.

rectifier An electrical device used to convert alternating current to direct current. Also, see *diode*.

regulator An electrical device used to control the output of a generator or alternator by controlling the current and voltage.

relay An electromagnetic switch. A low-current–operated control device used to open and close a high-current circuit.

resistance That property of an electrical circuit that tends to reduce the flow of current. Resistance is given in ohms; see *ohm*.

resistor A resistance device that, when installed in an electrical circuit, provides a reduction in current flow. Also, see *rheostat*.

revolutions per minute (RPM) The number of times a rotating member turns in 1 minute.

rheostat A variable resistance device used to regulate the flow of current in a circuit.

right-hand rule   A method used to determine the polarity of a coil. Also, see *left-hand rule*.

ROM   Read only memory.

rotor   That part of an alternator that rotates. The rotor of a motor is also referred to as an *armature*.

RPM   An abbreviation for revolutions per minute.

secondary circuit   The circuit that is connected to the secondary of a transformer.

selenium rectifier   A wafer-type rectifier, now seldom used, to convert alternating current to direct current.

semiconductor   A material that falls midway between good conductors and poor conductors (see also, *germanium* and *silicon*). The devices, diodes, transistors, silicon control rectifiers, triacs, and the like that are made up from semiconductor materials are called semiconductors.

series   See *series circuit*.

series circuit   Two or more components electrically connected end-to-end, so that the positive pole of one connects to the negative pole of the next.

short   The grounding, intentional or unintentional, of a current-carrying conductor or device.

short circuit   See *short*.

shorted winding   The grounding of the winding of a field, armature, or coil. The short may also be turn-to-turn instead of to ground. Also see *centrifugal short* and *short*.

shunt   A parallel circuit used to divide, usually in unequal percentages, the amount of current flowing from one point to the other through each circuit.

silicon   A semiconductor material.

silicon diode   A diode made from doped silicon.

slip ring   That part of an alternator rotor that provides a race through which the positive and negative brushes make electrical contact with the rotor.

solenoid   A coil around a movable magnetic core. The core becomes magnetized and moves when the coil is electrically energized.

solid-state   A term used to describe an electrical device which operates without mechanical function.

spark   A flow of electricity (electrons) through the air between two electrodes or contact points.

specific gravity   The weight of a solution with reference to water (H_2O), and which has an assigned value of 1.0.

splice   The joining of two or more conductors at a single point by use of fasteners or other devices.

static electricity   A charge of electricity which remains at rest until a suitable path is provided for its discharge.

stator   That part of the alternator that is stationary around the rotor. In motors, referred to as a *field*.

strobe light   A term often used for stroboscope.

stroboscope   An electrical neon lamp used to precisely time the ignition of an automobile engine.

substitute   To replace a part suspected of a defect with one of known quality.

sulfating   The formation of the inactive coating of lead (Pb) on the surface of battery plates; due to inactivity or nonuse.

sulfuric acid   A heavy, corrosive, high-boiling, liquid acid that is colorless. Mixed with distilled water, it forms the electrolyte used in storage batteries.

switch   A mechanical device used to open or close a circuit.

synchronize   To cause two or more events to occur at the proper time with respect to each other.

tachometer   An instrument used to measure the revolutions per minute of a rotating member, such as an engine.

temperature switch   An electrical switch device that is activated by temperature or a change in temperature.

terminal   A device attached to the end of a wire, cable, or device used to make an electric connection.

thermistor   A resistor with the characteristic of changing its value, in ohms, with a change of temperature. Also, see *resistor.*

thermostat   A heat-sensitive mechanical device found in the cooling system and used to regulate engine-coolant flow; an electrical switch actuated by heat.

three phase   The electrical power system in which voltages are generated 120 electrical degrees out of phase with each other.

time-delay relay   A relay with a timed circuit so that the points do not open (or close, depending on application) until a predetermined time interval after power is available.

tolerance   An allowable or permitted variation in dimensions.

torque   A twisting or turning effort measured in foot-pounds (ft-lb) or inch-pounds (in-lb).

transducer   A device that changes one form of energy to another.

transistor   A solid-state semiconductor that is a current amplifier or switch.

undercut   To cut away mica from between the commutator bars of an armature.

undercutter   A device (tool) used to undercut armatures.

vacuum   A pressure that is below that of atmospheric pressure.

volt   A unit of measure of electromotive force.

voltage   The electrical pressure that causes current to flow in a closed circuit.

voltage applied   The actual voltage read on a voltmeter at a given point in a circuit.

voltage available   The voltage delivered by a power supply.

voltage drop   The net difference in electrical pressure when measured on both sides of resistance (load). The loss of voltage caused by circuit resistance.

voltage regulator   See *regulator.*

voltmeter   An electrical instrument used to measure voltage.

watt   A unit of the measurement of electrical power. Amperage times voltage equals wattage.

watt-hour   A unit of electrical energy calculated by multiplying watts by hours.

wattmeter   An electrical instrument used to measure watts.

waveform   A graphical presentation of an electrical signal. Waveforms are normally seen on an oscilloscope. They are sometimes presented on an oscillograph.

zener diode   A diode that conducts current in the normal manner in the forward direction but breaks down at a predetermined voltage in the reverse direction. Its main use is as a voltage regulator.

Index

ELECTRICAL SYMBOLS COMMON TO AIR-CONDITIONING SYSTEMS

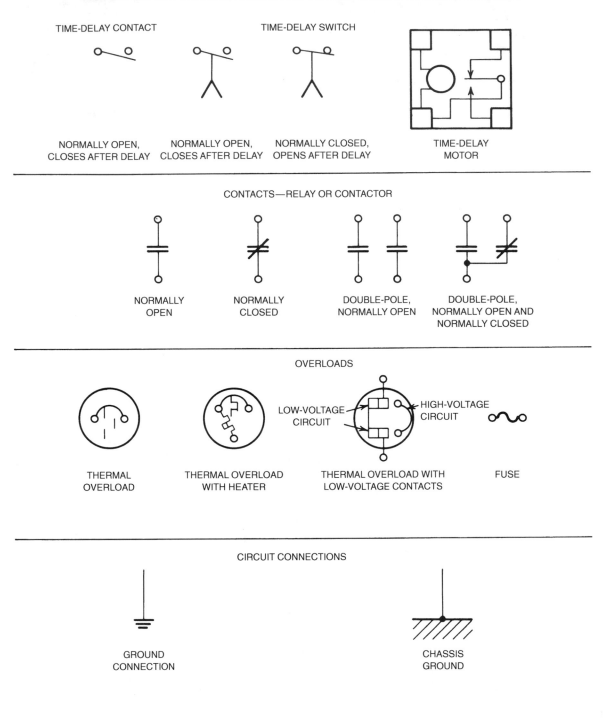

TIME-DELAY CONTACT

TIME-DELAY SWITCH

NORMALLY OPEN, CLOSES AFTER DELAY

NORMALLY OPEN, CLOSES AFTER DELAY

NORMALLY CLOSED, OPENS AFTER DELAY

TIME-DELAY MOTOR

CONTACTS—RELAY OR CONTACTOR

NORMALLY OPEN

NORMALLY CLOSED

DOUBLE-POLE, NORMALLY OPEN

DOUBLE-POLE, NORMALLY OPEN AND NORMALLY CLOSED

OVERLOADS

LOW-VOLTAGE CIRCUIT

HIGH-VOLTAGE CIRCUIT

THERMAL OVERLOAD

THERMAL OVERLOAD WITH HEATER

THERMAL OVERLOAD WITH LOW-VOLTAGE CONTACTS

FUSE

CIRCUIT CONNECTIONS

GROUND CONNECTION

CHASSIS GROUND

ELECTRICAL SYMBOLS COMMON TO AIR-CONDITIONING SYSTEMS

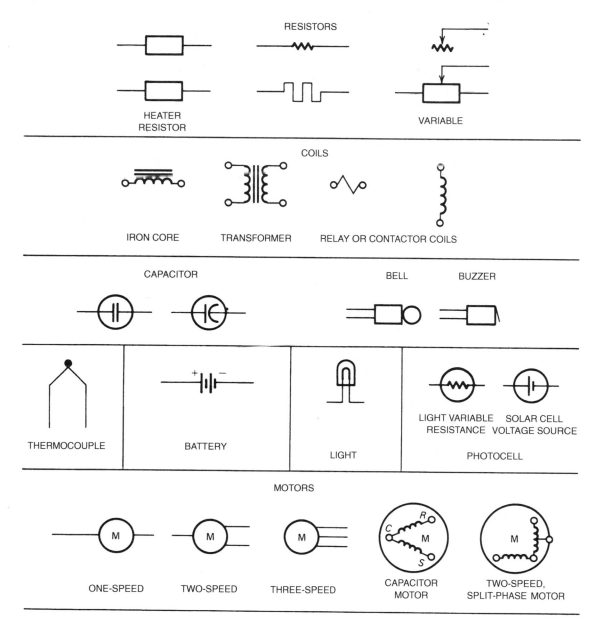

ELECTRICAL SYMBOLS COMMON TO AIR-CONDITIONING SYSTEMS

THREE-PHASE MOTOR CONNECTIONS

THREE-PHASE
WYE

THREE-PHASE
DELTA

SEMICONDUCTORS

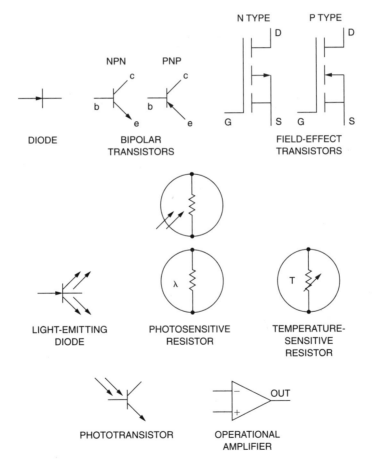

DIODE

NPN   PNP

BIPOLAR
TRANSISTORS

N TYPE   P TYPE

FIELD-EFFECT
TRANSISTORS

LIGHT-EMITTING
DIODE

PHOTOSENSITIVE
RESISTOR

TEMPERATURE-
SENSITIVE
RESISTOR

PHOTOTRANSISTOR

OPERATIONAL
AMPLIFIER